Advanced Techniques for
Embedded Systems Design and Test

T0180765

Advanced Techniques for
Embedded Systems Design and Test

edited by

Juan Carlos López
Technical University of Madrid

Román Hermida
Complutense University of Madrid

Walter Geisselhardt
Gerhard-Mercator-University Duisburg

KLUWER ACADEMIC PUBLISHERS
BOSTON / DORDRECHT / LONDON

A C.I.P. Catalogue record for this book is available from the Library of Congress.

ISBN 978-1-4419-5031-4

Published by Kluwer Academic Publishers,
P.O. Box 17, 3300 AA Dordrecht, The Netherlands.

Sold and distributed in North, Central and South America
by Kluwer Academic Publishers,
101 Philip Drive, Norwell, MA 02061, U.S.A.

In all other countries, sold and distributed
by Kluwer Academic Publishers,
P.O. Box 322, 3300 AH Dordrecht, The Netherlands.

Printed on acid-free paper

CONTENTS

CONTRIBUTORS

António Adrego de Rocha
University of Aveiro
Aveiro, Portugal

Paul Bisgambiglia
University of Corsica
Corte, France

Erwin R. Bonsma
University of Twente
Enschede, The Netherlands

Dominique Federici
University of Corsica
Corte, France

António B. Ferrari
University of Aveiro
Aveiro, Portugal

Milagros Fernández
Complutense Univ. of Madrid
Madrid, Spain

Víctor Fernández
University of Cantabria
Santander, Spain

Walter Geisselhardt
Gerhard-Mercator University
Duisburg, Germany

Sabih H. Gerez
University of Twente
Enschede, The Netherlands

Horst Günther
GMD
St. Augustin, Germany

Sonia M. Heemstra de Groot
University of Twente
Enschede, The Netherlands

Marc J.M. Heijligers
Eindhoven Univ. of Technology
The Netherlands

Román Hermida
Complutense Univ. of Madrid
Madrid, Spain

Heinz-Dieter Huemmer
Gerhard-Mercator University
Duisburg, Germany

Margarida F. Jacome
University of Texas
Austin, TX, USA

Juan Carlos López
Technical Univ. of Madrid
Madrid, Spain

María Luisa López-Vallejo
Technical Univ. of Madrid
Madrid, Spain

Hortensia Mecha
Complutense Univ. of Madrid
Madrid, Spain

José Manuel Mendías
Complutense Univ. of Madrid
Madrid, Spain

Eduard Moser
Robert BOSCH GmbH
Stuttgart, Germany

Paul G. Plöger
GMD
St. Augustin, Germany

Pablo Sánchez
University of Cantabria
Santander, Spain

Jean-François Santucci
University of Corsica
Corte, France

Valery Sklyarov
University of Aveiro
Aveiro, Portugal

Maite Veiga
University of Cantabria
Santander, Spain

Eugenio Villar
University of Cantabria
Santander, Spain

PREFACE

As electronic technology reaches the point where complex systems can be integrated on a single chip, and higher degrees of performance can be achieved at lower costs by means of putting together well-coupled hardware and software components, designers must devise new ways to undertake the laborious task of coping with the numerous, and non-trivial, problems that arise during the conception of such systems. On the other hand, shorter design cycles (so that electronic products can fit into shrinking market windows) put companies, and consequently designers, under pressure in a race to obtain reliable products in the minimum period of time.

Given this scenario, *automation* has been one of the answers of the design community. With the aid of highly sophisticated tools, designers have been able to deal with more and more complex designs, relying on computer programs to accomplish the more tedious tasks. *Abstraction* has also been traditionally used to allow the designers to get rid of annoying implementation details while better focusing on the decision making process. New methodologies have appeared supported by these two aspects. Their importance has been crucial in making it possible for *system designers* to take over the traditional electronic design process.

Embedded systems is one of the fields that these methodologies are mainly targeting. The inherent complexity of these systems, with hardware and software components that usually execute concurrently, and the very tight cost and performance constraints, make them specially suitable to introduce higher levels of abstraction and automation, so as to allow the designer to tackle better the many problems that appear during their design.

Hardware-Software Co-design is currently considered as the best method to deal with the design of those heterogeneous systems under the above discussed constraints, assuring design quality and reliability. Co-design explores reasonable implementations of a specific system by combining hardware and software solutions. The co-design process must evaluate the numerous implementation alternatives and the many tradeoffs involved in the system design. But the design space for such complex systems is huge and the quality of the final solution strongly depends on the way that space exploration has been carried out. Performance and cost estimations turn out to be essential in this task.

System specification gains importance in a co-design methodology. It helps the designer to better capture the conceptual view of the system with the minimum effort, that is, to describe the system functionality far away from implementation details (hardware and software parts, for instance). New languages offer more powerful means of expressing more abstract concepts in an easier way, facilitating the whole specification task.

On the other hand, the task of deciding which system functionality will be performed by either hardware or software components, becomes essential. The boundary between hardware and software represents the main tradeoff when designing a complex system, since that decision clearly affects the cost and performance of the final implementation. This is not an easy task and numerous efforts are being made to come up with new models and algorithms to perform the hardware-software partitioning.

Of course, the design of the hardware part is still a problem where more mature (software) methodologies are being applied. The *describe-and-synthesize* paradigm has been successfully used to take on this issue. Synthesis is a translation process from a behavioral description into a net of components that implements the given behavior. The ability to accept specifications with increasing abstraction level and the optimization of the synthesized results, as well as the search for new application domains, have been among the relevant topics in this field. Synthesis has been strongly marked for the last ten years at least by research on High-Level Synthesis, which has shaped a new specific activity field.

High-level synthesis has presented a polymorphic nature. The number of design related problems where these techniques are applied has grown considerably in the last years. There has been, of course, research on the classical problems of scheduling, allocation, binding, etc., but some other important topics have also been addressed. Area and delay estimation for example, has deserved a lot of attention because the high-level design decisions can only be correctly made if based on fine estimations of their impact on the final circuit layout.

Formal support for correct synthesis is an emerging area of activity, because the original promise of *correct by construction* synthesis results cannot be held true if we consider the enormous complexity of the synthesis tools. Therefore, there is a trend to link synthesis and verification so that the elementary decisions taken by synthesis algorithms can be verified, before continuing the design process.

After generating a design, traditional test approaches modify the design by adding new circuits that allow system functionality testing through the application of selected stimuli. Such modifications can imply significant increments of the initial chip area. However, recent research has shown that integrating testability considerations within the synthesis process may lead to testable designs as the whole circuit area is optimized.

The complexity of test generation has been traditionally tackled using divide-and-conquer strategies. But the use of more abstract views of the system has been introduced as a powerful way to cope with this problem. On the other hand, generation of stimuli is becoming influenced by new results derived from the analysis of the initial behavioral specification.

The design of complete systems on a chip is more and more performed by means of both, (numerous) reusable blocks taken from a library, and (a few) others specifically designed. All these components either include built-in self-test capabilities or come with test sets to exercise both structure and functionality. Behavioral specifications help to integrate the tests of the different components into an overall test procedure.

This book gathers almost all these topics, presenting recent developments in methodologies and tools for the specification, synthesis, verification, and test of embedded systems, characterized by the use of high-level languages as a road to productivity.

The use of new languages for the specification of these complex systems opens new ways to better describe (and analyze) the system functionality. Chapter 1 is devoted to this phase of the design cycle, studying the features that a good specification language for embedded systems should show and presenting the role it plays in a co-design methodology. Special emphasis on the use of ADA as embedded system specification language is given.

Early and efficient design space exploration is one of the main goals of the design community. Design assistance turns out to be a real need when the complexity of actual systems grows and the variety of potential implementations is huge. In Chapter 2, new ideas on modeling and characterizing design methodologies to support the search for cost-effective solutions to a given design problem are discussed. Algorithms, models and mechanisms for helping in the decision making process are proposed.

Hardware-Software partitioning is addressed in Chapter 3. The decision on the particular implementation of the different functional components has a strong impact on the cost and performance of the final product. Therefore, the way this process is accomplished (models, algorithms, cost functions,...) is of great importance. Furthermore, this decision is specially hard when it has to be based on imprecise and usually uncertain information. Knowledge-based techniques and fuzzy models are the foundation on which the partitioning system proposed in this chapter is built.

To test the value of co-design, the development of an industrial case study, based on a specific methodology, is analyzed in Chapter 4. The main steps of the design flow are studied in detail, describing methods, tools and results that help to draw some conclusions about the feasibility of using a co-design methodology in an industrial environment.

To assure the correctness of the design process has been one of the challenges of recent research works. Within a trend lately suggested by some authors, Chapter 5 discusses the possibility of incorporating formal verification actions during synthesis steps. Some ideas are presented about the formalisms that can be valid to support such actions, so that the two basic goals of correctness and efficiency can be achieved, within the specific scene of high-level synthesis.

Most of today's electronic systems are characterized by really tight timing constraints. Good scheduling of the involved operations is necessary to assure the fulfillment of such constraints. Chapter 6 is focused on the study of overlapped scheduling techniques, aimed at exploiting parallelism in loops execution, guaranteeing an optimum utilization of the available time slots.

The use of the object-oriented model to the design of reconfigurable controllers is addressed in Chapter 7. Although control design reusability was formerly a clear reason for the interest of this topic, the advent of dynamically reconfigurable field programmable devices, makes it even more attractive. The proposed approach considers, starting from behavioral specifications under the form of graph schemes and their varieties, the modification or extension of the controller functionality, as well as the reuse of control components.

High-level estimation of circuit physical features is another topic that has attracted considerable attention in the last years in the hardware synthesis field. Chapter 8 presents an approach that shows how a detailed knowledge of the physical design style of a specific tool, for a given technology, can be successfully applied to the creation of fast accurate estimators, which can be used when high-level design decisions have to be made.

Chapter 9 is devoted to automatic creation of testable hardware structures during the synthesis process, always starting from Hardware Description Languages. Following an outline of the design process and its relation to testing, the principal techniques of Design for Testability are analyzed and the current state-of-the-art of RT level, as well as high-level, test synthesis are described. System-level test considerations are also addressed.

What can be gained from using behavioral system descriptions in automatic test pattern generation (ATPG) is the objective of Chapter 10. In general, the hierarchy of abstraction encompasses three levels, the structural at gate and switch level, the functional and the behavioral. Characteristics of ATPG at these levels are described together with the main algorithms in use. Exploiting the behavioral view in structural ATPG leads to a hierarchical approach which combines high fault coverage at the structural level with the efficiency of behavioral pattern generation.

Behavioral fault simulation is addressed in Chapter 11. As it has already mentioned, behavioral descriptions are used to cope with complexity in design and test generation. But a prerequisite for test generation at any level (structural, functional or behavioral) is the existence of an efficient fault simulator at every respective level. After giving the motivation of behavioral testing and describing the main concepts in behavioral test pattern generation, the benefits of coupling a behavioral fault simulator with a deterministic and a pseudo-random behavioral test pattern generator are explained.

Every chapter is biased by the work performed at each participant laboratory, but includes a general description of the state-of-the-art in every addressed topic. Therefore, the reader can find in this book a good way to know recent developments on the design and test of complex heterogeneous systems, as well as to have an overview of the different issues that, in this specific field, currently concentrate the attention of designers and CAD tools developers.

Acknowledgments

The idea of this book was conceived within a Network of Laboratories that, under the project *"Behavioral Design Methodologies for Digital Systems"* (BELSIGN), have been cooperating since 1995, with the funding of the Human Capital and Mobility Programme of the European Union. Most of the research that is reported here has been carried out in the different labs of the network, and has received the beneficial influence of the other teams through discussions, meetings, visits and interchanges.

The editors would like to express their gratitude to all the authors for the efforts to promptly deliver their originals, as well as for their enthusiastic promotion of this project. The support and encouragement received from Professor Francisco Tirado, the network coordinator, must also be recognized. Special thanks must be given to the European Commission for the financial support to our activities and to Kluwer Academic Publishers through its representative Mike Casey, who was always receptive to our project.

Juan Carlos López
Román Hermida
Walter Geisselhardt

1 EMBEDDED SYSTEM SPECIFICATION

Eugenio Villar
Maite Veiga

1.1 Introduction

Electronic Design Automation (EDA) has evolved dramatically during the last ten years covering almost satisfactorily all the design steps from the RT-level description down to the final implementation. Recently, the first commercial behavioral synthesis tools have appeared in the market. Nevertheless, electronic system complexity supported by current microelectronic technology leads to the need for more sophisticated and powerful design methodologies and tools. This need will become greater in the near future as technology continues to evolve. As a consequence, computer-aided, system-level design methodologies represent the next step in EDA tool generation still to be developed. This is the reason for the recently increasing interest in Electronic System Design Automation (ESDA) tools.

By complex electronic systems, we mean systems composed of one or several PCBs, implementing analog and digital functions and interacting with their environment through analog and digital signals, sensors and actuators.

Design methodologies for such systems are far from maturity, particularly, regarding mixed-signal specification, analog/digital partitioning, analog design and mixed-signal verification. After analog/digital partitioning, both the analog subsystem specification and the digital subsystem specification are decided.

In this chapter, we will focus on the specification of complex digital systems, particularly, embedded systems. Embedded systems can be defined as computing and control systems dedicated to a certain application [Mic94]. In the most general

1

J.C. López et al. (eds.), Advanced Techniques for Embedded Systems Design and Test, 1-30.
© 1998 *Kluwer Academic Publishers.*

case, an embedded system will be composed of different components implementing hardware (HW) and software (SW) functions in different ways.

Hardware functions may be implemented using off-the-shelf components, programmable logic and Application-Specific Integrated Circuits (ASICs). Software functions may be implemented in several ways. So, off-the-shelf processors, both general-purpose processors or application-specific processors such as Digital Signal Processors (DSPs) can be used. Also, Application Specific Standard Processors (ASSPs), that is, commercial processors used as a core inside an ASIC. As a more customized option, Application Specific Integrated Processors (ASIPs), that is, processors designed specifically for the application, can be considered.

The behavior of an embedded system is defined by its interaction with the environment in which it operates [GVN94]. In many cases, they have to react to the environment by executing functions in response to specific input stimuli. These systems are called reactive systems [Mic94]. When the system functions must execute under predefined timing constraints, they are called real-time systems [Mic94]. Apart from general purpose computing systems, most current electronic systems are reactive, real-time systems. They are found in a wide variety of applications such as the automotive field, manufacturing industry, aerospace, telecommunication, consumer electronics, etc.

Although design of embedded systems comprising hardware and software functions has been a common practice since the first microprocessors were developed, classical approaches are no longer valid to ensure high quality designs under strict time-to-market and design cost constraints. There are several reasons leading to this situation. First, the increasing complexity of today's microprocessors and associated real time software. Second, the increasing complexity of the associated hardware performing all those tasks which owing to timing constraints or cost can not be allocated to the microprocessor. Third, the increasing complexity of the hardware/software interface implementing the hardware/software communication mechanisms needed in these complex real-time applications. Fourth, the increasing integration capability opening the way to ASSP-based designs, or even, ASIP-based designs. As a consequence, development of new, efficient design methodologies for such complex embedded systems currently constitutes one of the most important challenges. These design methodologies are called hardware/software co-design methodologies and the specification methods for them constitute the main objective of this chapter. From now on, the word 'system' will be used for digital embedded systems and 'system specification' for hardware/software (HW/SW) co-specification of such systems.

1.2 Specification Requirements

Although both software development and hardware design used almost matured methodologies, co-design of systems composed of software and hardware functions currently constitutes an open unsolved problem. System specification represents one of the most important challenges in embedded system design. For the first time, digital design is faced as a unified problem for which all the implementation possibilities commented previously can be considered. In particular, software and hardware functions become implementation alternatives for the same functional specification. This fact imposes a general requirement on the specification method which must be capable of specifying in a natural way computational algorithms which are going to be partitioned and their parts implemented either in hardware or software. Additional, more concrete requirements are the following [GVN94][GuL97].

1.2.1 Programming Constructs

In most cases, the behavior of an embedded system can best be described using a sequential algorithm. In fact, system design can be seen essentially as a programming activity [GuL97]. System specification by means of a programming language is particularly natural for those functions to be implemented in software. Additionally, behavioral synthesis has made this kind of description natural for functions to be implemented in hardware. It is sometimes argued that sequential algorithms are well suited for specifying control-dominated systems but they have severe limitations when dealing with data-dominated systems. By control-dominated systems we mean systems with a low degree of parallelism, whose behavior is easily represented by states and state transitions. By data-dominated systems we mean systems with a high degree of parallelism, whose behavior is easily represented by data flow. In HW/SW co-design, this distinction is superfluous. Hardware low-level parallelism found in data-dominated systems has to be determined during behavioral synthesis and, therefore, should not be imposed at the system-level specification. The specification method has to be able to specify in a natural way the algorithm independently of whether it is going to be implemented in software, in hardware with a low degree of parallelism or in fully pipelined hardware. This is possible because the algorithm is independent from its final implementation. Digital Signal Processing (DSP) is a good example for this statement. So, for instance, a Fast Fourier Transform can either be implemented on a microprocessor, a DSP processor or in hardware depending on the throughput required. It is an objective of any co-design tool to be able to identify the best implementation in each case from the same specification.

1.2.2 State Transitions

In some cases, embedded systems are best conceptualized as having different states[1] of behavior. Each state may represent any arbitrary computation. In most cases, after state computation completion, the transition condition is evaluated and the next state decided. Figure 1.1 shows an example of such a description for a monitoring system.

---

[1] The term "state" is used here as a computation mode of an algorithm and, therefore, in a more general sense than the state in a Finite State Machine (FSM).

This description means is particularly natural for reactive systems [Cal93]. It can be easily translated to a programming language code for execution and analysis purposes using similar description styles for FSMs in VHDL.

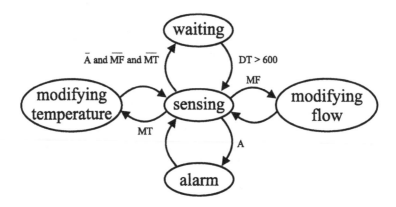

Figure 1.1. State transition specification for monitoring system

1.2.3 Partition and Hierarchy

Partition and hierarchy represent two of the most important means of handling complexity. Following the principle of "dividing and conquering", a problem which is too complex to be understood as a whole is decomposed into smaller pieces which are decomposed again until manageable components are found. The problem is then approached by ensuring that each component is behaving correctly and its composition makes the corresponding component in the higher level in the hierarchy behave correctly too. Two kinds of hierarchical partitions are desirable in the embedded system specification means, structural and behavioral.

Structural partition and hierarchy here refers to the capability of decomposing the system into a set of interconnected components which may have their own internal structure. As each component has its own behavior, structural partition is closely related to concurrency which will be commented afterwards. If a programming language is used, structural partition is supported if concurrent tasks are supported.

Behavioral partition and hierarchy here refers to the capability of decomposing the system behavior into distinct subbehaviors which may have their own internal behavioral partition and hierarchy. Almost all programming languages support behavioral partition and hierarchy by procedures and functions.

In general, embedded system specification will require a mixture of both structural and behavioral hierarchical partition as shown in Figure 1.2.

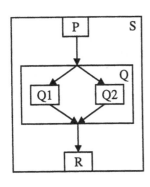

```
procedure S is
   procedure P;
   procedure Q is
      task Q1;
      task Q2;
      •
      •
      task body Q1 is
      begin
         •
         •
      end Q1;
      task body Q2 is
      begin
         •
      end Q2;
   begin
   end Q;
   procedure R;
      •
begin
   P;
   Q;
   R;
end S;
```

a) Hierarchical partitioning.   b) ADA specification.

Figure 1.2. System S

1.2.4 Behavioral Completion

In some cases, the overall behavior of the system can be better when it, and the subbehaviors into which it is divided reach a completion point. At that point, any computation in the behavior has been executed and all the variables updated. Behavioral completion helps designers to conceptualize any behavior at any hierarchical level as an independent module facilitating its verification inside the complete system. Additionally, behavioral completion facilitates the hierarchical partitioning of the complete specification.

1.2.5 Concurrency

In most cases, complex behavior can be better conceptualized as a set of concurrent subbehaviors. Concurrency is an important requirement in any specification method in order to support structural partitioning. Concurrency can be applied with different granularity. In a HW/SW co-specification, the thickest grain granularity is represented by task-level concurrency. Any behavior can be partitioned into different tasks such as procedure Q in system S of Figure 1.2 which has been divided into tasks Q1 and Q2. A finer grain granularity is represented by statement-level concurrency. The most classic example of this level of concurrency is represented by the parallel, pipelined implementation of the sequence of statements in a loop.

An even finer grain granularity would be the operation-level concurrency in which, provided the data dependencies are satisfied, two operations can be performed sequentially or in parallel. So, for instance, in the sequence of statements of Figure 1.3, operations O1 and O2 can be performed either sequentially or in parallel.

```
v1 := x1 + x2;        -- O1
v2 := x1 - x2;        -- O2
v3 := v1 * v2;
v4 := v3 + x3;
```

Figure 1.3. Operation level concurrency

The finest grain granularity would be the bit-level concurrency which is decided at the logic level. So, for example, an N-bit addition can be implemented either sequentially, leading to a ripple-carry adder or concurrently, leading to a carry-look-ahead adder.

A tradeoff is always possible between sequential and concurrent implementations. Sequential implementations are, in general, slower but cheaper (in terms of number of gates and power consumption) while, concurrent implementations are, in general, faster but more expensive. Any synthesis tool at any level should be able to perform sequential-concurrent transformations exploring the design space for solutions satisfying the design requirements. So, bit-level concurrency is decided during logic optimization in RT-logic synthesis. Operation and statement-level concurrency are decided during scheduling in behavioral synthesis. Task-level concurrency is decided during partitioning in HW/SW co-design.

During partitioning, the functions to be implemented in hardware and the function to be implemented in software are decided. Both kinds of will be executed concurrently in the final implementation. If a single-processor architecture is used, software functions will be compiled for serial execution. If a multi-processor architecture is used, software tasks will be allocated to the available processors and executed concurrently. Hardware tasks will be implemented as one or several hardware components operating concurrently. Software engineering indicates that the specification language should reflect the structure of the application domain. If the application domain contains inherent parallelism, then the design and realization of the software product will be less error-prone, easier to verify and easier to adapt if concurrency is available in the specification language [BuW95]. As the embedded will contain concurrent operation, concurrency is an important feature of any HW/SW co-design specification means. Sequential computations are deterministic and may hide non-deterministic behavioral modes which may appear in the final implementation. A concurrent specification may help to detect them and if necessary, to avoid them.

Moreover, concurrency leads to more flexible specifications. By serialization, from a concurrent specification, several sequential specifications can be obtained[2]. Any of these specifications will correspond to a particular execution mode of the original concurrent one. From this point of view, it is clear that a sequential specification implies a certain degree of over-specification. Although parallelism extraction from a sequential specification is always possible, it is usually a much more complex task than serialization.

1.2.6 Communication

Concurrent tasks need to interchange information in order to perform the complete computation of the system. Information interchange can be done by different communication mechanisms which can be grouped in two different models, shared-memory communication and message-passing communication as shown in Figure 1.4.

In the shared memory communication model, a shared storage medium is used to interchange data. As it will be described in more detail afterwards, the safe use of shared data is not straightforward and some mechanisms have to be used in order to ensure mutual exclusion. By mutual exclusion we mean the prevention of concurrent access to the same data. A shared medium is said to be persistent when any data written by one task in certain positions remains intact until the same or another task rewrites new data in the same positions. A memory is persistent by default. A shared medium is said to be non-persistent when any data written by one task is lost when the same or another task reads the data. Examples of non-persistent media are stacks and buses. The shared memory model allows easy implementation of broadcasting, that is, making any data generated by a task accessible to any other task.

In the message passing communication model, a channel is used to interchange data. A channel is an abstract entity to be decided later on. It can be implemented as a memory, a FIFO, a bus, etc. In each of the communicating tasks, send/receive mechanisms have to be defined. A communication channel can be uni-directional, if one task always sends and the other always receives, or bi-directional, if both tasks may send and receive. Depending on the tasks using the channel, it can be point-to-point, if only two tasks are connected through the channel or multiway, if more than two tasks use it. The message-passing communication through the channel is said to be blocking, if the sending task has to wait until the receiving task reaches the point in which it is prepared to receive the data. Blocking communication implies

---

[2] In this section, sequential and concurrent specifications are compared. Of course, any concurrent behavior can be emulated using a sequential algorithm. Nevertheless, this requires the addition of an execution kernel which makes sense for emulation or simulation purposes but not for system specification.

synchronization between tasks and can be used exclusively for this purpose without any data transfer. The advantage of blocking communication is that it does not require any additional storage medium. As a disadvantage, suspending task execution may imply a degradation in system performance. The communication is said to be non-blocking if the sending task sends the data independently of the state of the receiving task which at some point will read the data. Non-blocking communication requires some additional storage medium to store the data until they are read, a FIFO usually, but, it normally results in a better system performance. If, for any reason, the sending and/or receiving speeds are different, the storage medium may become full or empty, thus requiring blocking the data transfer to avoid losing data.

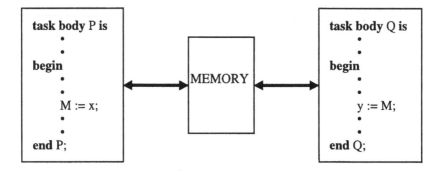

a) Shared memory.

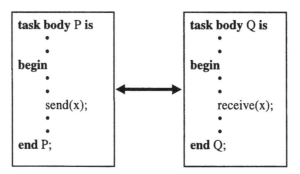

b) Message passing.

Figure 1.4. Communication models

1.2.7 Synchronization

In many cases, concurrent tasks need to adapt their own execution to the execution state of other tasks or external events. Synchronization refers to the mechanisms used to suspend the execution of a task until another task reaches a certain state or an external event occurs. There are two main synchronization mechanisms, control-dependent and data-dependent. In control-dependent synchronization, task synchronization is defined directly by the control structure of the specification. An example of such a synchronization mechanism was shown in Figure 1.2. Based on ADA task execution semantics, both tasks Q1 and Q2 have to finish their execution in order for procedure Q to finish and pass control to procedure R. In data-dependent synchronization, any of the two communication mechanisms described above are used. In shared-memory synchronization, one task suspends until another writes an appropriate value in the shared storage medium. As was commented before, message passing synchronization was possible using blocking communication.

1.2.8 Exception Handling

In some cases, the occurrence of an execution condition or an external event requires a task to suspend its computation immediately and pass control to a predetermined task. Abnormal computation conditions (i.e. division by 0) and hardware or software interruptions are examples of high-level exceptions. In some cases it is necessary to abort the computation being performed by a certain task and execute a different computation instead. The corresponding triggering mechanism would represent a high-level reset. The hardware reset represents an RT-logic exception.

1.2.9 Non-determinism

Non-determinism refers to the quality of a system to react in an unpredictable way to a certain event. Two kinds of non-deterministic behavior can be found when an embedded system is specified, functional non-determinism and temporal non-determinism. Functional non-determinism refers to the possibility of behaving in different ways under the same sequence of input patterns. As commented before, a concurrent specification is non-deterministic by nature as different results can be obtained depending on the order of execution of tasks and thus, reflects more accurately any non-deterministic behavior in the final implementation. In the general case, system behavior should be predictable, yielding the same results from the same inputs [GuL97]. Thus, functional non-determinism, constitutes an essential feature at the system specification level in order to detect it as early as possible and, if necessary, to avoid it. This kind of non-determinism will be described in more detail afterwards when the communication among tasks is analyzed. Temporal non-determinism refers to not knowing the time spent by a certain computation. This is also an essential feature, as at the system-level exact timing is not yet known. It can vary dramatically depending on the final partition and implementation of both the hardware and the software parts. As will be commented afterwards, only functional and/or domain timing restrictions in terms of minimum and/or maximum timing constraints among data are relevant. Coming back again to the example of Figure 1.2, it is impossible to know in advance the execution time of any of the procedures P, Q and R. Moreover, in most cases, it will be impossible to know which of the two tasks Q1 or Q2 will finish earlier.

1.2.10 Timing

By timing, the information about the temporal behavior of the system at a certain abstraction level is understood. Timing information is one of the most important characteristics of the system being designed and, in fact, it can be used to conceptually identify the different levels of abstraction.

Once the system has been implemented, it is possible to measure its exact temporal behavior[3]. Before fabrication, once all the components of the system as well as their physical place on the board or chip have been decided, a very precise logic model can be generated providing simulation results with very accurate timing. Before placement and routing, a logic-level model is available including the typical delays of the logic components. At the RT level, all the output assignments and register transfers in each clock cycle are known and a clock-based simulation of the circuit can be done. At the algorithmic level, only the causal relationship between events is known and only some estimations about the temporal behavior of both the hardware and the software functions can be provided. At the specification stage, no temporal information about the system is available and only the timing constraints to be imposed are defined.

Depending on whether the constraint applies to external or internal conditions, two kinds of timing constraints can be distinguished, functional and domain timing constraints. By functional constraints we mean temporal restrictions to the behavior of the system or module being designed. It is an objective of the design process to ensure that the final implementation of the system performs the computations at a speed satisfying the functional constraints. By domain constraints we mean temporal characteristics on the inputs of the system or module. It is an objective of the design process to ensure that the final implementation of the system is able to function correctly under the domain constraints imposed by the environment in which it has to operate.

Timing constraints are typically specified as a pair of values corresponding to the minimum and maximum timing values allowed. When both values are equal, an exact timing constraint is specified. When the constraint applies to events on different signals, it is called a delay. Delays can be specified as a certain amount of time or a number of clock cycles. When the constraint applies to events on the same signal, it is called a bit rate. A bit rate can be specified as data/time, bauds

---

[3] This temporal behavior is unique in the sense that not even another implementation of the same system will behave in exactly the same way. At the same time, it may vary with temperature and may change throughout the life of the system.

(bit/second) or frequency. Figure 1.5 shows an example of the different timing constraints which can be imposed on a system or module specification.

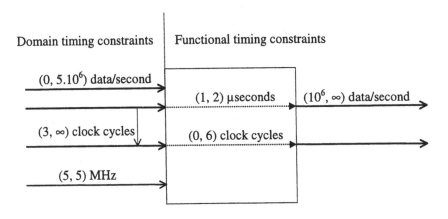

Figure 1.5. Timing constraints

Certain timing constraints may be impossible to satisfy. Timing constraints have to be analyzed for consistency in order to ensure that an implementation exists satisfying them [KuM92].

1.2.11 Object-Orientation

Object-oriented programming is a very useful feature for writing reusable components that are easily customized and assembled into large specifications. Besides facilitating the development of new specifications, object-oriented programming makes it easier to modify existing specifications [Coh96]. This presents in turn two additional advantages. The first being to facilitate the development of complex specifications particularly in a team-based design project. The second, facilitating the maintainability of previously developed specifications. Object-oriented programming is currently a usual programming approach. Its extension to hardware design is currently under investigation based on an object-oriented extension of VHDL [NeS96].

1.3 Specification Languages

Circuit description at the RT-level evolved from an immature period in the 70s characterized by a very large number of different HDLs proposed and developed to the current mature situation characterized by the usage of a few standard HDLs (basically VHDL and Verilog). At the system level, the situation is similar to that during the 70s at the RT-level: Many languages are used in different HW-SW co-design methodologies based on different underlying models [GVN94][Ram93]. Among them, the following are the most important.

1.3.1 Application-Specific Languages

Specific languages are being used for specific applications. So, for instance, SDL and LOTOS are CCITT standard languages for telecommunication system specification. Silage is specialized in the specification of digital signal processing hardware systems. Esterel is based on the perfect synchrony hypothesis which represents a formalization of the clock-based (RT) paradigm being developed for the specification of reactive systems. Moreover, new application-specific languages are still being proposed [PuV97].

1.3.2 State-Oriented Models

Several graphical notations have been proposed for system specification based on the state transition paradigm. One of the most classical is represented by Petri Nets and their extensions [GVN94][Ram93]. A Petri Net is a bi-graph of places and transitions. The state of the system is represented by the tokens the places contain. State transitions are determined by transitions firing. A transition can fire when all its input places contain at least one token. When a transition fires, a token is removed from each of the input places and a token is put in each of the output places.

Another classical graphical specification notation is represented by StateCharts [GVN94][Ram93]. StateCharts are an extension of the FSM model in order to facilitate the description of complex behaviors. A further extension is represented by the SpecCharts which allow the association of any arbitrary behavior to any state [GVN94].

All these models can be translated to a programming language for execution purposes. Their main advantage comes as graphical specification entries. This kind of specification means is particularly suited for reactive systems.

1.3.3 C-based Languages

C is one of the most widely used languages for system specification. The reason for this derives more from its popularity as a programming language and the corresponding availability of compilers than from its features as a system specification language. In fact, C has several limitations when used in this kind of application, particularly, its lack of support for concurrency. Its use as a system specification language imposes specific styles on the user [GuL97]. In order to overcome some of these limitations, several C-based specification languages have been proposed such as HardwareC [GuM93] and C^x [ErH92]. Nevertheless, their spread and support are limited. JAVA is being considered as an alternative in this sense. Nevertheless, JAVA has been defined specially for distributed systems and its application to real-time system specification would require of an improvement to the language.

1.3.4 VHDL

For a hardware designer, the extension of VHDL up to system specification represents a very attractive approach as both the language knowledge and many of the tools can be re-used. Nevertheless, VHDL has severe limitations in the specification of complex systems performing real-time functions to be implemented in hardware or in software depending on the design constraints. Nevertheless, on the contrary the case of C, VHDL has severe limitations as a system specification language. The most important limitation of VHDL derives from the underlying, event-driven, logic simulation mechanism it is based on. At the system-level, exact timing is not yet known. It can vary dramatically depending on the final partition and implementation of both the hardware and the software parts. Only functional and/or domain timing restrictions in terms of minimum and/or maximum timing constraints among data are relevant. VHDL on the contrary, imposes exact timing which, in most cases, requires an unaffordable over-specification. If simulation timing is not taken into account, simulation results may become unreliable as they may depend on a particular event synchronization. Moreover, the simulation time is unnecessarily large [CSV92]. Concurrency is another essential feature supported by VHDL. Nevertheless, although data concurrency is fully supported by VHDL, control concurrency is only partially supported. Moreover, the communication mechanism is extremely simple. It is based on synchronous shared memory (signals). Although VHDL'93 also supports asynchronous shared memory based on shared variables, very few commercial simulators fully implement this feature. As commented before, functional non-determinism is an essential feature at the system specification level. Only through global variables is it possible to make a VHDL description non-deterministic.

1.3.5 ADA

In this chapter, ADA is proposed for the specification of complex, real-time, embedded systems containing functions to be implemented either in hardware or software. Its suitability for this kind of application will be analyzed. Its role in a complete HW-SW co-design methodology will be described. ADA can overcome most of the limitations found in the above-mentioned languages and notations. ADA, as a programming language, is executed without any exact timing which accelerates the verification and debugging of the system specification. ADA has been specially defined as a concurrent language for real-time system design. In fact, several specification and design methodologies for embedded real-time systems are based on this language [Cal93][PuV97]. It contains several communication and synchronization mechanisms among tasks. In order to support particular real-time constraints, it contains some time clauses. In an ADA-VHDL co-simulation environment, timing restrictions could be annotated and taken into account. Additionally exceptions are supported.

ADA compilation allows the generation of the executable software code which is difficult in other languages like VHDL. As commented before, from the designer point of view, the system specification and RT languages, if not the same, should be as similar as possible. Translation from ADA to VHDL for hardware synthesis is made easier by the syntactical similarities between both languages. In addition to the characteristics commented so far, ADA supports object orientation which represents an additional advantage in the specification of complex systems.

1.4 Co-Design Methodology

The co-design methodology proposed is based on the interrelated use of both ADA and VHDL. Both are standard languages supported by the IEEE with a very similar syntax and based on the same programming concepts. In fact, the ternary ADA/VHDL/VHDL-AMS represents a coherent set of languages able to cover most of the design tasks from the specification of mixed analog-digital systems down to the final implementation. The proposed co-design methodology for the digital subsystem is that shown in Figure 1.6.

From the ADA specification, the HW/SW partition is done and the corresponding HW and SW algorithms are decided. The software algorithm will be extracted directly from the ADA specification. By compilation, the executable program will be obtained ready for storage in the program memory for execution by the processor in the target architecture. The hardware algorithm will be obtained from the ADA specification and translated to VHDL. Syntactical similarities between ADA and VHDL make this translation process easier. From the VHDL description the corresponding hardware will be obtained by behavioral synthesis.

Some restrictions have to be imposed on the ADA specification part to be implemented in hardware. The majority of them will derive from the VHDL synthesis syntax and semantics of the behavioral synthesis tool. Regarding the ADA specification, three different situations arise. The first, is represented by those ADA features directly translatable to VHDL. These features do not represent any problem and due the syntactical similarities between ADA and VHDL only minor modifications will be needed. The second, is represented by those ADA features without any direct translation to behavioral VHDL but, which with some modifications can be converted to VHDL code supported by the behavioral synthesis tool. These features will be called indirectly supported. The third, is represented by those ADA features without possible translation to behavioral VHDL. Any ADA specification module containing this kind of unsupported features does not admit hardware implementation and has to be moved to software during partitioning.

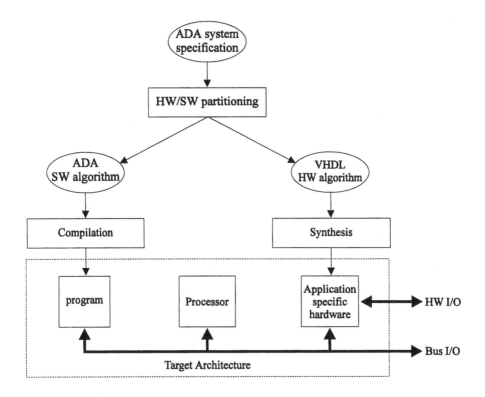

Figure 1.6. Proposed co-design methodology

1.5 ADA System Specification

The aim of the system specification methodology is to permit the complete specification of embedded systems in ADA through a series of concurrent tasks and procedures, without introducing any type of restrictions on the language usage. The main advantage of this general approach is its flexibility. No syntactical constraint is imposed. Nevertheless, synthesis efficiency is always related to the capability of the synthesis tool to identify the implementation associated with the behavior described. This requires an implementation semantics in the specification used [ViS93]. As a consequence, there will be some parts of the code which due to the style in which they have been specified could lead to unsuitable or inefficient implementations. In HW/SW co-design this is particularly probable in those tasks to be assigned to hardware which when described using computationally complex algorithms (i.e. code handling complex data structures, recursive functions, etc.) could lead to unsuitable implementations. It is a designer responsibility to use suitable description styles for the whole specification and, particularly, for those functions with a high probability to be allocated to hardware. In fact, an important aspect in ADA system specification is the definition of appropriate description styles.

Specification code generation with appropriate description style refers to both the syntactical as well as the semantic aspects of the code. On the one hand, a certain code may not admit any hardware implementation for syntactical reasons. On the other hand, a certain code can be syntactically supported by the compiler or the hardware synthesis tool but leading to inefficient implementations.

1.5.1 Specification Structure

The system specification is conceived as an entity with inputs and outputs as shown in Figure 1.7.a. In order to verify the specification under the operational conditions in which the system has to function, an environment or test-bench has to be developed. This environment has to produce the inputs to verify the system and has to store the outputs produced by the system as shown in Figure 1.7.b.

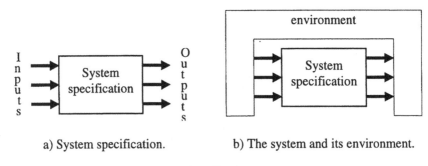

a) System specification.   b) The system and its environment.

Figure 1.7. Specification structure

As both the system specification and its environment have to operate concurrently, they have to be developed as ADA tasks activated by the main subprogram. When dealing with large programs, the ADA separate compilation facilities should be exploited. By separate compilation, writing a program in several parts and submitting these parts to a compiler one at a time is meant. Separate compilation makes it possible to change one part of a program without recompiling it entirely; allows part of a program being developed by different programmers to be kept in independent files; and allows the program to be written, checked for syntactic errors and tested piece by piece. Separate compilation in ADA does not mean independent compilation. While compiling one unit, an ADA compilation system uses information in other compilation units to perform the same consistency checks between separately compiled parts of a program that are performed between different parts of a single compilation unit. This allows the early discovery of errors that might otherwise not be detected until the separately compiled units of a large program are combined [Coh96]. Using ADA separate compilation, the system specification structure would be that shown in Figure 1.8.

The entries of task "specification" allow easy identification of the inputs and outputs of the embedded system. They play a similar role to ports in a VHDL entity. While task "specification" is to be totally or partially designed, task "environment" is not. Thus, the above-commented description style for implementability applies for the former. For the latter, only the general recommendations to be followed when programming in ADA apply.

```
procedure system_name is
    -- inputs declaration
    -- outputs declaration
    task specification is
        entry inputs ();
        entry outputs ();
        pragma map_module (inputs, protocols.RS232);
    end specification;
    task body specification is separate;
    task environment;
    task body environment is separate;
begin
    null;
end system_name;
```

a) ADA main procedure.

```
separate (system_name)              with gnat.io; use gnat.io;
task body specification is          separate (system_name)
    -- declarative part             task body environment is
begin                                   -- declarative part
    -- sequence of statements       begin
end specification;                      -- sequence of statements
                                    end environment;
```

b) Specification task.   c) Environment task.

Figure 1.8. System specification in separate units

ADA is an imperative, high-level language supporting a wide variety of programming facilities. In this Section only the most relevant characteristics of the language and their implications in embedded system specification will be introduced. For a more detailed description of the language, visit [BuW95] and [Coh96].

1.5.2 Partition and Hierarchy

Partition and hierarchy is supported in ADA through subprograms, tasks and blocks.

1.5.2.1 Subprograms

ADA subprograms are quite similar to VHDL subprograms both in their usage and syntax. As commented before, subprograms represent the way to support behavioral partition and hierarchy. Except for recursion and overloading which will be commented afterwards, they do not represent any problem in HW/SW co-design.

1.5.2.1.1 Recursive Subprograms

Like VHDL, ADA supports recursion. A subprogram is recursive directly or indirectly. A subprogram is said to be directly recursive when it contains a call to itself. A subprogram is said to be indirectly recursive when it contains a call to a subprogram containing a call to the original subprogram. In this case both subprograms are recursive and are said to be mutually recursive.

As recursion is not supported by synthesis tools, and no translation to an equivalent synthesizable code exists, it represents an unsupported hardware feature.

1.5.2.1.2 Overloading

Overloading in ADA corresponds to the same concept and has the same syntax as in VHDL. Overloading of predefined functions (with a certain hardware semantics) is not allowed by synthesis tools and therefore, it represents an unsupported feature.

1.5.2.2 Blocks

As in VHDL, ADA allows the encapsulation of statements in blocks. While in VHDL, blocks contain concurrent statements, in ADA, blocks contain sequential statements. As an equivalent procedure can be obtained from any block and procedures are directly supported, the blocks are also directly supported. The block may have its own declarative part. The scope of any variable declared in the declarative part of a block extends only from the declaration to the end of the block. Figure 1.9 shows an example of a block in its simplest form for calculating the factorial "Ft" of a positive number "N".

```
Factorial: declare
               Fn : Integer := 1;
           begin
              Ft := 1;
              while Fn < N loop
                 Fn := Fn + 1;
                 Ft := Ft * Fn;
              end loop;
           end Factorial;
```

Figure 1.9. Block for factorial computation

1.5.2.3 Tasks

As it will commented later in more detail, tasks represent a key concept in order to support concurrency at the specification level. If allocated to hardware, tasks will be translated to VHDL as processes. Additional VHDL code will be generated in order to adequately implement the required communication between the task and its environment. Also the software part will require, in general, additional code with the same purpose.

1.5.3 Exceptions

Reliable systems should work properly even if abnormal computation modes are reached due to malformed or unconsidered input data, unconsidered execution modes or hardware malfunctions. The specification should avoid such unexpected situations but, in order to be reliable, it should account and be prepared for them. Such situations are called exceptions in ADA and the language provides mechanisms for dealing with them. When an unexpected situation occurs, it is said that an exception is raised. Responding to the exception is called handling the exception.

Exceptions can be implemented in hardware by hardware conditions forcing the circuit to jump to specific states to execute the corresponding computation and eventually completing the task. As they do not have any direct equivalency in VHDL but, additional VHDL code can be generated to implement them, they represent an indirectly supported feature.

1.5.3.1 Predefined Exceptions

There are many ADA rules which are checked at run-time. When any of those ADA rules is violated during program execution, a predefined exception is raised.

1.5.3.2 Propagation of Exceptions

When a certain statement causes an exception to be raised, the sequence of statements of the block, the subprogram body or the package to which the statement belongs is abandoned and the handler for that exception, if any, is executed instead of the remainder of the sequence. This completes the execution of the block, the subprogram or the package initialization. If no handler for the exception is found, the execution of the sequence of statements is abandoned anyway and the exception is propagated to the higher level of control; in the case of a block statement, to the sequence of statements containing the block; in the case of a subprogram, to the point where the subprogram was called. If the exception is propagated to the main subprogram and no handler is found, the execution of the program is aborted. In the case of a package body, if no handler is found, the exception causes a fatal error in initializing the program.

1.5.3.3 Exception Handling

As stated before, exceptions should be handled in order to ensure correct continuation of the program or, at least, controlled termination of the program thus increasing its reliability. So, for instance, in the block of Figure 1.9, there is an unlikely possibility for product overflow. In order to prevent the program from stopping execution, a handler can be included. In the example, if a product overflow is produced, the factorial is put to "0" indicating that contingency but the program continues its execution:

```
Factorial: declare
              Fn : Integer := 1;
           begin
              Ft := 1;
              while Fn < N loop
                 Fn := Fn + 1;
                 Ft := Ft * Fn;
              end loop;
           exception
              when Constraint_Error =>
                 Ft := 0;
           end Factorial;
```

Figure 1.10. Block for factorial computation including an exception handler

1.5.3.4 User-defined Exceptions

The programmer can define his/her own exceptions when an improper computation mode is detected which should be treated in the same way as a predefined exception. Suppose, for example, that instead of a block, the factorial computation of Figure 1.10 is to be performed by function Factorial and that, for some reason, operand N may be negative or null. In order to deal with such a contingency, a user-defined exception Negative_Number is defined. In this case, when a non-positive number is encounter, an exception is raised and propagated to the subprogram where the function was called:

```
Negative_Number: exception;
...
function Factorial (N: Integer) return Integer is
     Fn, Ft : Integer := 1;
   begin
   if N <= 0 then
      raise Negative_Number;
   else
      while Fn < N loop
         Fn := Fn + 1;
         Ft := Ft * Fn;
      end loop;
      return Ft;
   end if;
   exception
      when Constraint_Error =>
         return 0;
      when Negative_Number =>
         raise;

   end Factorial;
```

1.5.4 Interrupt Handling

An interrupt is an event, usually generated outside the processor executing the program which suspends the normal execution of the program, executes a specific code in response to the interrupt and, when finished, continues the execution of the program at the point it was suspended.

Interrupts can be explicitly described in the system specification. These interrupts will affect the whole system. During HW/SW partitioning, the corresponding interrupt handlers can be assigned either to SW or HW. As in the case of exceptions, they do not have direct translation to VHDL but, as they can be implemented in hardware, they represent an indirectly supported feature. As will be commented later, low-level interrupts will be required in order to implement the HW/SW communication mechanisms. The ADA support for interrupt handling and low-level programming are required in this case.

The standard ADA mechanisms for handling interrupts are provided by compilers which conform to the System Programming Annex. In this Annex, package Ada.Interrupts is defined. This package provides the types:

```
type Interrupt_Id is implemetation_defined;
type Parameterless_Handler is access protected procedure;
```

Interrupts can be declared as expressions of type Interrupt_Id. Interrupt handlers can be installed as parameter-less protected procedures of type Parameterless_Handler. This ensure interrupt blocking as a protected procedure cannot be called while an operation on the protected object is in progress. An interrupt handler can be installed using the pragma Attach_Handler:

```
pragma Attach_Handler (Handler, Interrupt);
```

The following is an example of a simple interrupt handler:

```
protected Handler_Example is
   procedure Input_Handler;
   pragma Attach_Handler (Input_Handler, Input);
private
   -- private resources
end Handler_Example;

protected body Handler_Example is
   procedure Input_Handler is
   begin
      -- Interrupt code
   end Input_Handler;
end Handler_Example;
```

1.5.5 Concurrency

As commented previously, concurrency represents an essential system feature to be captured during system specification. Concurrency is supported in ADA through tasks. Once created, a task will perform its computation until completion.

1.5.5.1 Task Types

Task types allow us to declare tasks which can be created afterwards. The values of the task type will correspond to tasks. Once activated, these tasks will perform their own computation concurrently. The full declaration of a task type consists of its declaration and body. The basic syntax for a task declaration is the following:

```
task_type_declaration ::=
    task type identifier [discriminant_part] [is task_def];

task_def ::=
    { entry_declaration }
    [ private { entry_declaration } ]
end [identifier];
```

The basic syntax for a task body is the following:

```
task_body ::=
    task body identifier is
        declarative_part
    begin
        sequence_of_statements
    end [identifier];
```

The discriminant part defines the discrete or access parameters that can be passed to any of the tasks of the type when they are created. Task entries may be public or private. Private entries are invisible to the task type user.

The following is an example of a task type declaration of users of a lift system:

```
task type User (Start_Floor, Destination_Floor: Natural);

task body User is
    declarative_part
begin
    sequence_of_statements
end User;
```

When a single task is to be declared, no task type is needed. In this case, the syntax for task declaration is simpler:

```
single_task_declaration ::=
    task identifier [discriminant_part] [is task_def];
```

Task specification and environment in Figure 1.8 are examples of single task declarations.

1.5.5.2 Task Creation

A task will be created by declaring objects of a task type. Values should be given to the discriminants, if any. The behavior of the task will be that of the task type body. The following will be the declaration of two users in floor 3 willing to go to floor 6:

```
User1, User2 : User(3, 6);
```

Tasks can be created dynamically providing an access type to the task type is declared:

```
type User_Pointer is access User;
Task_Pointer : User_Pointer;
```

Now, new User tasks can be created dynamically during program execution:

```
Task_Pointer := new User(5, 0);
```

The name of the task is Task_Pointer.**all** and it corresponds to a user on floor 5 willing to go to floor 0.

A pointer to a task created by an object declaration can be obtained provided the task was created as aliased. So, for instance, if the following user is created:

```
User3 : aliased User(0, 6);
```

a pointer to task User3 may be used:

```
type User_Pointer is access all User;
Task_Pointer : User_Pointer;
.
.
Task_Pointer := User3'Access;
```

After the assignment, task User3 can be referenced by its name or by Task_Pointer.

1.5.6 Communication

Tasks can communicate among themselves through two main mechanisms, shared variables and rendezvous. Only these two simple communication mechanisms have to be implemented during interface synthesis when an specification task and an environment task are involved or when one of the tasks is assigned to software and the other to hardware. Based on them, any other, more complex communication mechanism can be developed (channels, broadcasting, etc.)

1.5.6.1 Shared Variables

If a variable is visible to two tasks, both of them can read and/or modify it. This is the most simple task communication mechanism. Nevertheless, communication through shared variables may lead to functional non-determinism. As commented previously, it is essential to detect any functional non-determinism in the specification in order to, if necessary, provide the mechanisms to avoid it in the final implementation. Two kinds of functional non-determinism can be distinguished. The first, which we will call unexpected non-determinism, is caused by the concurrent access to the variable by the two tasks. Suppose, for instance, the example in Figure 1.11. In this example, tasks T1 and T2 are counting independent events (i.e. persons entering a building through different gates) but, which have to be accumulated in the same variable X. Even a simple operation like "X + 1" generally requires of several clock cycles in the final implementation independently of whether one or both tasks are implemented in hardware or software, as shown in Figure 1.11 b). If both tasks access the variable in an actual concurrent way the final result may be unexpectedly incorrect.

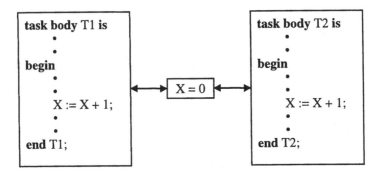

a) Shared memory access specification.

R1 <= X;	.
R1 <= R1 + 1;	R2 <= X;
X <= R1;	R2 <= R2 + 1;
.	X <= R2;

$$X = 1$$

b) Shared memory access implementation and result.

Figure 1.11. Unexpected functional non-determinism

This kind of non-deterministic behavior can be avoided by the use of protected objects and a correct implementation procedure for them. So, Figure 1.12 shows how to avoid the functional non-determinism of Figure 1.11 using a protected object. A more detailed explanation of protected objects is outside the scope of this Chapter and can be found in [BuW95] and [Coh96].

The second kind of functional non-determinism, which we will call essential non-determinism, derives directly from the concurrent nature of the system implementation and the consequent unpredictable order of execution of tasks. As an example, consider again the specification in Figure 1.11 in the case of task T2, instead of incrementing X, multiply it by 2. If task T1 is executed first, the result would be '1'. If T2 is executed first, the result would be '0' instead. This kind of non-determinism can not be avoided even ensuring in the final implementation mutual exclusion for task access to shared variables and, therefore, has to be detected as earlier as possible in order to, if necessary, avoid it.

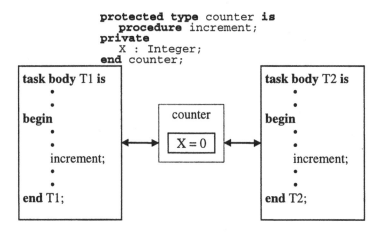

Figure 1.12. Protected shared memory access specification

1.5.6.2 Rendezvous

Rendezvous constitutes the built-in, message-passing, communication mechanism in ADA. The rendezvous is a direct communication mechanism in which a task calls an entry in another task and the latter accepts the call. Then, data are moved from the calling task to the accepting task through its input parameters and in the opposite direction through the output parameters. The task which reaches the call or the accept statement first waits for the other. Once both tasks reach the call and the accept statements, the rendezvous can start. Then, the actual values are copied into the corresponding input formal parameters.

The code inside the accept statement is executed. Finally, the actual values are copied into the corresponding output formal parameters. The rendezvous finish and both the calling and the accepting tasks can continue their own concurrent computation.

As a general recommendation, the computation inside an accept statement should be kept as simple as possible in order to allow the calling task to continue its computation as soon as possible.

The rendezvous is performed between two and only two tasks. Nevertheless, several tasks may call the same entry of a task. In that case, the calls are queued and accepted one at a time. In the ADA Real-Time System annex, two queuing policies are defined, the FIFO_Queuing and the Priority_Queuing. The first one is the default queuing policy in which calls are served in a first-in-first-out basis. The latter, orders calls by task priorities and can be specified using the pragma:

```
pragma Queuing_Policy (Priority_Queuing);
```

A task with several entries may wait simultaneously for calls to any of them and accept the call which arrives first. This is the general case in an embedded system with several inputs. ADA supports this possibility with the selective accept statement as shown in Figure 1.13.

```
task specification is
    entry Input1 (data: in Integer; CTS: out Bit);
    entry Input2 (data: in Integer; CTS: out Bit);
end specification;

task body specification is
    •
    •
    select
        accept Input1 (data: in Integer; CTS: out Bit) do
            Din1 := data;
            CTS := full(1);
        end Input1;
    or
        accept Input2 (data: in Integer; CTS: out Bit) do
            Din2 := data;
            CTS := full(2);
        end Input2;
    end select;
    •
    •
end specification;
```

Figure 1.13. Selective accept statement

Acceptance of a call to an entry may be conditional. This feature, which allows us to dynamically open or close select alternatives, is supported in ADA by guards. A task calling an entry in a closed select alternative will be forced to wait until it is open. So, for instance, in the previous example, when in a certain input no more data can be accepted, the environment is simply informed through the CTS signal. This does not prevent the environment continuing to send data. If for any reason, the corresponding input should be closed, a guard can be used as shown in Figure 1.14.

```
task body specification is
   .
   .
   select
      when full(1) = '0' =>
         accept Input1 (data: in Integer; CTS: out Bit) do
            Din1 := data;
            Din1_Stored := Din1_Stored + 1;
            if Din1_Stored = Din1_Limit then
               full(1) := '1';
            end if;
            CTS := full(1);
         end Input1;
   or
      when full(2) = '0' =>
         accept Input2 (data: in Integer; CTS: out Bit) do
            Din2 := data;
            Din2_Stored := Din2_Stored + 1;
            if Din2_Stored = Din2_Limit then
               full(2) := '1';
            end if;
            CTS := full(2);
         end Input2;
   end select;
   .
   .
end specification;
```

Figure 1.14. Selective accept statement with guards

Sometimes, the task has to perform some computations even in the case when no calls to any of the entries in a select alternative are produced during a certain amount of time. In this case, a delay alternative can be used:

```
select
   accept Input1 (data: in Integer; CTS: out Bit) do
      Din1 := data;
      CTS := full(1);
   end Input1;
or
   accept Input2 (data: in Integer; CTS: out Bit) do
      Din2 := data;
      CTS := full(2);
   end Input2;
or
   delay Waiting_Period;
end select;
```

Now, task specification will wait during the Waiting_Period for calls to any of the two entries. If no call is produced during this time, the execution of the select statement is finished.

In the case when the task can not wait for calls to any of the entries, a select statement with an else part can be used:

```
New_Data := True;
select
   accept Input1 (data: in Integer; CTS: out Bit) do
      Din1 := data;
      CTS := full(1);
   end Input1;
or
   accept Input2 (data: in Integer; CTS: out Bit) do
      Din2 := data;
      CTS := full(2);
   end Input2;
else
   New_Data := False;
end select;
```

In this example, if no call to any of the two inputs can be accepted immediately, the else part is executed and the execution of the select statement is completed.

The delay alternative and the else part represent incompatible means for a task to react when no calls are produced to its entries. As a consequence only one of the two can be used in a selective accept statement.

The selective accept statement allows control over when a task should wait for calls on its entries and when to accept them. In a similar way, a calling task can also get control over the way to proceed if the call is not accepted. So, a timed entry call allows a task to stop waiting if the call is not accepted during a certain amount of time:

```
select
   T2.Input(data);
or
   delay Waiting_Period;
end select;
```

In the case when the task can not wait for the call to be accepted, a conditional entry call can be used:

```
select
   T2.Input(data);
   Data_sent := True;
else
   Data_sent := False;
end select;
```

1.5.7 Synchronization

As commented previously, any of the communication mechanisms described so far can be used for synchronization purposes. Nevertheless, the simplest way in ADA for synchronizing to tasks, that is, to force the two tasks to reach a certain computation state and then allow them to proceed concurrently again, is an accept statement without body as shown in Figure 1.15. In order to continue execution, both tasks have to meet at the synchronization point.

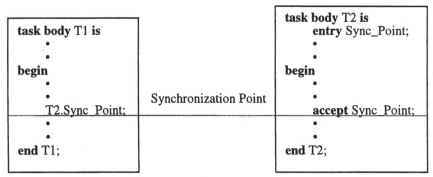

Figure 1.15. Task synchronization

Sometimes, the execution of a task has to be aborted as a result of an event in another task. In a system specification, this can refer to an external event if the task causing the abortion belongs to the environment or an internal event if it is a task belonging to the specification itself. Abortion represents an additional exception mechanism in the general sense commented in Section 1.2.8. In fact, it represents a high-level reset mechanism. This feature is supported in ADA by the asynchronous select:

```
select
   T1.reset;
   Reset_Computation;
then abort
   Normal_Computation;
end select;
```

Procedure Normal_Computation represents the abortable computation performed by the task under normal conditions. If the call to entry reset of task T1 is accepted, procedure Normal_Computation is aborted and procedure Reset_Computation is executed instead. The call to entry reset of task T1 is called the triggering statement. The triggering statement can also be a call to a protected entry or a delay statement:

```
select
   delay until Max_Execution_Time;
   Reset_Computation;
then abort
   Normal_Computation;
end select;
```

In this case, if the computation is not completed in its maximum execution time, the computation is aborted.

A task can abort an entire task with an abort statement:

```
abort T;
```

This statement provokes task T to terminate.

1.6 Conclusions

This chapter has dealt with the specification of embedded systems. The requirements that any system specification method should fulfill have been analyzed, and the main languages used for embedded system specification have been described. The limitations found when using these languages have been commented on. ADA has been proposed, and has been shown powerful enough, for the specification of complex systems containing functions to be implemented either in hardware or in software. It is particularly suited for real-time systems specification and design. Its syntactical similarity to VHDL facilitates the development of a HW-SW co-design methodology using VHDL for hardware specification and design.

References

[BuW95] A. Burns and A. Wellings: *"Concurrency in ADA"*, Cambridge University Press, 1995.

[Cal93] J. P. Calvez: *"Embedded real-time systems: A specification and design methodology"*, Wiley, 1993.

[Coh96] N.H. Cohen: *"ADA as a second language"*, McGraw-Hill, 1996.

[CSV92] E. Casino; P. Sánchez and E. Villar: "A high-level VHDL simulator", *proc. of Spring VIUF'92*, May, 1992.

[ErH92] Ernst, R. and Henkel, J.: "Hardware-software co-design of embedded controlers based on hardware extraction", *proc. of the IEEE International Workshop on HW/SW Co-design*, September, 1992.

[GuL97] R.K. Gupta and S. Liao: *"Using a programming language for digital system design"*, IEEE Design and Test of Computers, April-June, 1997.

[GuM93] R.K. Gupta and G. de Micheli: *"Hardware-software co-synthesis for digital systems"*, IEEE Design and Test of Computers, September, 1993.

[GVN94] D. D. Gajski, F. Vahid, S. Narayan and J. Gong: *"Specification and design of embedded systems"*, Prentice-Hall, 1994.

[KuM92] D.C. Ku and G. de Micheli: *"High-level synthesis of ASICs under timing and synchronization constraints"*, Kluwer, 1992.

[Mic94] G. de Micheli: *"Synthesis and optimization of digital circuits"*, McGraw-Hill, 1994.

[NeS96] W. Nebel and G. Schumacher: "Object-Oriented hardware modeling - Where to apply and what are the objects?", *proc. of EuroDAC'96 with EuroVHDL'96*, IEEE, September, 1996.

[PuV97] P.H.A. van der Putten and J.P.M. Voeten: *"Specification of Reactive hardware/software systems"*, CIP-Data Library Technishe Universiteit Eindhoven, 1997.

[Ram93] F. Rammig: *"System level design"*, in "Fundamentals and standards in hardware description languages", edited by J. Mermet, Kluwer, 1993.

[ViS93] E. Villar and P. Sánchez: *"Synthesis applications of VHDL"*, in "Fundamentals and standards in hardware description languages", edited by J. Mermet, Kluwer, 1993.

2 SUPPORTING EARLY SYSTEM-LEVEL DESIGN SPACE EXPLORATION IN THE DEEP SUBMICRON ERA

Margarida F. Jacome
Juan Carlos López

2.1 Introduction

In this chapter we consider the problem of assisting designers in selecting cost-effective *system-on-chip architectures* for high-volume communications, automotive control, video processing, and consumer electronics products. Efficient solutions for such products, in terms of performance and power consumption, have traditionally been obtained by designing application-specific integrated circuits (ASICs) fully customized to the application's requirements. However, with the recent advent of *application-specific instruction-set processors* (ASIPs or IPs), this scenario is rapidly changing [MiSa96]. An *ASIP*, or *core*, is a programmable processor whose architecture and instruction set are customized to specific classes of applications [Goos96]. Compared to general-purpose processors, the architectural specialization of an ASIP results in better area/performance and power/performance ratios.

The trend in embedded systems technology is thus to integrate, on the same IC, one or more ASIPs (or, for that matter, any other programmable processor or micro-controller), memory blocks, peripheral blocks (such as A/D and D/A converters), and possibly a few custom (i.e., function-specific) hardware blocks. In such ICs, most of the system functionality is implemented in software. Custom hardware blocks, or "hardware accelerators", are used for time-critical tasks [MiSa96].

The size, complexity, and heterogeneity of such system-on-chip designs poses a number of challenging research problems. First, in order to allow for a systematic iterative search within the correspondingly huge and complex design space (during

31

J.C. López et al. (eds.), Advanced Techniques for Embedded Systems Design and Test, 31-52.

architecture selection), some form of *formal characterization* of that space, at a conveniently *high-level of abstraction*, is required. A second important problem is that of generating accurate (early) performance, area, and power consumption *estimates*, in order to reliably guide designers during the architecture-selection phase. Finally, the ability to define *design libraries* that can effectively support an aggressive pruning of the design space, is yet another critical research issue.

In this chapter we discuss our ongoing research aimed at developing a methodology for supporting a systematic search for cost-effective system-on-chip architectures, addressing the three issues referred to above. The remainder of this chapter is organized as follows. We start by giving an overview of our proposed framework for early, system-level design space exploration. The main methodological and modeling issues investigated in the context of this framework are then discussed in some detail. We first discuss how our design formalism is used to characterize the design space. Next we describe the internal algorithm and architectural models used to represent design decisions taken during the exploration process. Mechanisms for assessing candidate architectures are then proposed. The chapter concludes with a discussion of design libraries required for supporting the proposed methodology.

2.2 A Framework for Assisting Early System-level Design Space Exploration

In this section we give an overview of our proposed methodology for assisting designers in selecting cost-effective system-on-chip architectures for complex systems. Our approach assumes that the *initial system specification* is given by: (1) an algorithm-level description of the system behavior (written in C or VHDL), annotated with profiling information; and (2) a set of timing, area, and/or power requirements.

The final system architecture specification includes: (1) the definition of a set of *architectural components* and *interfaces*; (2) a first-cut *partition* of the system's algorithmic description into a set of algorithmic segments, each of which is to be implemented by one of the architectural components; (3) a *floorplan* for the system architecture; and (4) timing/area/power *budgets* for the various architectural components. Observe that architecture selection is the first step of the top-down design process. Accordingly, the architecture specifications produced during this design step are to be considered in more detail, and refined, during the following steps of the top-down design process.

Figure 2.1 shows our proposed framework for assisting designers in producing the architectural specifications characterized above. The framework has three major components: a *design space exploration* sub-system, an *estimation* sub-system, and a *library of algorithms and components*. In what follows we briefly discuss the role of these three components.

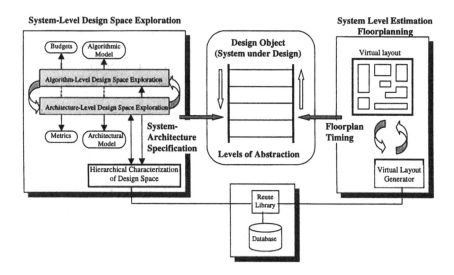

Figure 2.1. A Framework for System-Level Design Space Exploration

2.2.1 The Design Space Exploration Sub-system

The search for a cost-effective system-on-chip architecture is an iterative process where behavioral (i.e., algorithmic) and structural (i.e., architectural) *design issues*, or areas of design decision, should be synergistically considered, as symbolically shown in Figure 2.1.

In order to allow for a systematic, controlled *pruning* of the design space, the ordering according to which these various design issues are to be considered (or revised, in case of failure) should be established based on their level of pervasiveness with respect to the relevant figures of merit (e.g., on performance, silicon area, etc.). A *formal design space representation*, incorporated in the design space exploration sub-system, defines such design issues, as well as their ordering. This formal representation is constructed using the formalism of design to be discussed later on in this chapter.

As also indicated in Figure 2.1, two internal models are maintained by the exploration sub-system: (1) an *algorithmic model*, representing the system behavior; and (2) an *architectural model*, representing possible realizations of that behavior.

Using the algorithmic model, timing, area, and power *budgets* are first assigned to the various computations or tasks defined in the behavioral description of the system. Such assignments are based on: (1) the performance and other requirements stated in the initial system specification; and (2) actual measures on the relative complexity of these various computations/tasks. This initial budgeting is then incrementally refined during the design space exploration.

As alluded to above, the main goal of early design space exploration is to identify a system-on-chip architecture that can cost-effectively meet the system requirements (properly translated into budgets). Before dropping a less expensive architecture and considering a more expensive one, it is thus crucial to make sure that adequate system implementations for the first (less costly) architecture have been considered.

Algorithm substitutions and/or behavior preserving algorithmic transformations (such as parallelization of code segments or loop unrolling), as well as different input/output encodings, may be applied during the design space exploration, in an attempt to meet the required budgets, or re-distribute them in a more adequate way. For example, parallelization of a time-critical code segment may be tried in order to increase performance, at a possible cost of increased area. Moreover, different partitions and mappings of the system behavior on to the various architectural components should be considered during the design space exploration. For example, time-critical code segments may be moved from the ASIP(s) into the function-specific hardware component(s), once again with a possible increase in area.

Naturally, alternative solutions may realize different *overall* trade-offs (in area/performance and/or power/performance). Moreover, the set of possible solutions is huge. Thus, our approach provides a set of *abstract design metrics* aimed at helping designers to *quickly* converge to *"heuristically good" design points* -- i.e., points that will hopefully realize the best overall trade-offs for a given target architecture.

2.2.2 The Estimation Sub-system

After a promising design point for the system architecture under investigation is identified, *feasibility analysis* needs to be performed, i.e., hard numbers on the achievable performance (and/or area and power consumption) need to be *estimated*, in order to assess the solution's *potential* to meet the required budgets. This is the task of the frameworks' *estimation* subsystem.

In particular for deep submicron CMOS technologies, with most of the delay occurring in the interconnect, physical effects must be considered early in the design process. *Floorplanning,* the first and most critical step in the physical design, determines the spatial and interface characteristics of the architectural components of an IC, such that the desired physical and electrical constraints are satisfied.

Accordingly, the system-on-chip architecture specification produced by our framework also includes floorplanning. Specifically, a floorplan is created for the IC, and the budgets are assessed, based on a novel method that combines: (1) the top-down, "abstract" design decisions made using the first sub-subsystem; with (2) bottom up estimates generated from *"virtual layouts."* Note that, with a view at supporting deep-submicron designs, virtual layouts provide an *approximate physical implementation* of the system under design, using previous designs stored in the library of algorithms and components (see Figure 2.1).

2.2.3 The Library of Algorithms and Architectural Components

As discussed previously, the library of algorithms and components is designed to effectively support the two main sub-systems of the framework. Specifically, it supports the design space exploration sub-system by characterizing potential (alternative) algorithms, implementation styles, etc., in a way that allows for a quick, aggressive pruning of the design space. It supports the estimation subsystem, by supplying the previous designs needed to construct the virtual layouts, i.e., the physical design approximations of the IC under design.

2.3 A Formalism of Design for Characterizing the Design Space

As alluded to above, due to the typically huge size and complexity of the design space, its exploration cannot be undertaken in an ad hoc way. As will be seen, our approach formally characterizes the design space, i.e., the space of viable design alternatives for the application domains of interest, in a way that allows for its systematic exploration and pruning. This formal characterization is defined using the formalism of design introduced in [Jaco93][JaDi96], which is briefly summarized in this section.

2.3.1 Characterizing Design Objects

A *design object* is an abstraction of a physical device or process and is characterized in terms of a set of properties or features that describe its *structure*, i.e., the way it is realized, and its *behavior*, i.e., the way it should function. (An "Adder" is an example of a classes of design objects for the design discipline of digital VLSI circuits.) Design is viewed as the process of deriving a complete specification, i.e., a set of properties that uniquely characterizes a particular design object, starting with a sub-set of such properties (i.e., an initial specification).

The *specification definition* for a class of design objects, say "Adders", is thus the set of *property definitions* which are necessary for characterizing each design object in the particular class, i.e., each adder design. An example of one of such property definitions would be "Word Size."[1] A *property instance*, or simply a *property*, describes the feature designated by the property definition, e.g., 16 or 32 bits for "Word Size."

As alluded to above, the specification definition of a class of design objects can be partitioned into two general categories: behavioral specifications and structural specifications. Each of these specifications may be further partitioned into three sub-categories: *requirements, restrictions,* and *descriptions.*

---

[1] For simplicity, the examples are given informally.

2.3.1.1 Requirements

Behavioral and structural requirements are typically part of the initial system specification. They define the "givens" of the design problem, i.e., they are the properties that the object under design must meet when the design process is complete. Timing constraints (on latency or throughput) are examples of behavioral requirements. Examples of structural requirements are silicon area and power consumption.

2.3.1.2 Restrictions

Behavioral and structural restrictions are properties that are employed by the designer to prune the design space in order to reduce design complexity. Restrictions thus represent *design issues* or areas of design decision -- designers typically address restrictions during what is called the *conceptual design* phase of the design process. For example, structural restrictions allow the designer to constrain the design object's structure to specific topologies, layout styles, and fabrication technologies (e.g., a designer may wish to constrain the structure of a given digital circuit to a specific layout style, such as gate array, and to CMOS fabrication technologies).

2.3.1.3 Descriptions

Behavioral and structural descriptions are properties that *fully* (i.e., completely) define the behavior and the structure of a given design object, respectively. Behavioral descriptions indicate, via mathematical equations or behavioral description languages, how a design object reacts, or should react, to specific sets of stimuli applied to its inputs. Structural descriptions define the design object as an interconnection of a number of structural building blocks. In our framework, for example, the initial system specification includes a behavioral description (given in VHDL or C), fully describing the intended behavior of the system under design.

2.3.2 Design Object Decomposition

Because of its importance in controlling complexity, we now briefly discuss design object decompositions performed during the design process. Note that behavioral descriptions (relating the outputs of the design object to its inputs) can be specified using simply, and directly, the *fundamental* connectives and/or constructs of mathematical formalisms and/or behavioral description languages defined for a specific level of design abstraction. Alternatively, a behavioral description can be specified in terms of a set of *assumed* behavioral sub-descriptions. A behavioral sub-description is said to be "assumed if it is not explicitly represented using only fundamental connectives and/or constructs. An assumed behavioral sub-description can, thus, be seen as a *"non-primitive behavioral building block"*, for the particular abstraction level.

A behavioral description incorporates a *functional decomposition* in the sense that it expresses the (complex) behavior of an entire design object in terms of less complex (primitive or non-primitive) "behavioral building blocks." A behavioral description is said to incorporate a *fundamental functional decomposition* if it contains only "primitive behavioral building blocks", otherwise, is said to incorporate *a non-fundamental functional decomposition.*

Similarly, a structural description is an interconnection of a number of structural building blocks. Such building blocks can be either primitive building blocks, or non-primitive building blocks. A primitive building block cannot be expressed in terms of other, simpler, building blocks defined at the same abstraction level. Before the design process can commence, structural primitive building blocks must be available for each abstraction level of the specific design discipline. Furthermore, a behavioral description, must exist for each of the structural primitive building blocks. Such behavioral descriptions are called models. (Note that a primitive structural building block may have different models depending on the operational ranges in which it will be used and/or the required accuracy with which we want to reproduce the real physical process.)

As in the previous case, a structural description can also be seen as a structural decomposition. A structural description is said to incorporate a *fundamental structural decomposition* if it only uses structural primitive building blocks. Otherwise, if the structural description uses at least one structural non-primitive building block, it is said to incorporate a *non-fundamental structural decomposition.*

Given the above, we may say that two of the main objectives of our early system-level design space exploration framework are: (1) to create a non-fundamental structural decomposition of the system into a set of architectural components; and (2) to create a non-fundamental behavioral decomposition of the system behavior, done by *partitioning* its algorithmic description into a number of algorithmic segments (to be implemented by each of the previous components).

2.3.3 The Discipline Hierarchy

Typically, for each given discipline, it is possible to identify a number of different classes of design objects. For instance, for the discipline of VLSI Digital Circuits, we may design "ALUs", "Multipliers", and "Adders." On the basis of the notion of a discipline is the fact that different classes of design objects typically share important properties. For example, a design discipline may use the same levels of abstraction to represent several different classes of design objects. Furthermore, different classes of complex design objects may share important structural and/or behavioral "non-primitive building blocks". This suggests that it is important to properly organize the entire set of classes of design objects that constitute a given discipline. In this section we describe such an organization, called the discipline hierarchy.

We can formally define the *discipline hierarchy* as a two dimensional organization of ordered sets of property definitions, as discussed in the sequel. The two dimensions used to construct the discipline hierarchy are the abstraction dimension and the specialization dimension.

The *abstraction* dimension organizes the design hierarchy into abstraction layers. More specifically, the abstraction dimension partitions the specification definition of each individual class of design objects across the several (defined) levels of design abstraction. The resulting sub-specifications constitute what we call the *facets* of the class of design objects.

The principle behind the use of abstraction is to reduce the complexity of finding a design object (in a given class of design objects) that satisfies a given specification. This is accomplished by allowing the designer to begin by considering only those properties contained in the most abstract specification. When this initial specification is met by the object under design, the designer moves down in the abstraction hierarchy, and considers the next, lower level (i.e., more detailed) specification. Naturally, the properties encompassing the most pervasive design decisions should be "placed" at the higher levels of design abstraction. Examples of levels of abstraction for the discipline of VLSI digital circuits design are "register-transfer level", "logic level", "transistor level", and "physical level."

We now discuss the specialization dimension of the discipline hierarchy, which serves a dual purpose. The first purpose is obvious -- it allows for discriminating between the different classes of design objects which may coexist for the design discipline. The second purpose is somewhat more subtle -- it allows for expressing commonalties among the different classes of design objects defined for the particular design discipline. Because of its critical importance in terms of enabling a systematic and aggressive pruning of the design space, in what follows we discuss the specialization concept in more detail.

Specialization (or conversely, generalization) allows different classes of design objects to share property definitions -- this is done by allowing "specialized" classes of design objects to inherit common properties from more "general" classes of design objects. Note that, since a class of design objects is simply a collection of properties, generalization is realized at the property definition level, as discussed in the sequel. Specifically, generalized properties, collapsing families of alternatives, should precede more detailed (i.e., specific) properties, whenever that may be useful from the point of view of quickly identifying achievable ranges of performance, power consumption and other relevant figures of merit. As before, those increasingly specialized properties are organized into a hierarchy of correspondingly increasingly specialized facets (i.e., views of classes of design objects, at each level of design abstraction).

We feel that this notion of specialization will be increasingly critical if one wants to be able to effectively support any kind of systematic design space exploration of *large* design spaces. In particular, in order to allow for an aggressive pruning of the design space, groups/families of algorithms, implementation styles, etc., ought to be merged (i.e., "*generalized*") during the initial exploration phases. By doing so, they become selectable as a "viable family of design alternatives" or, conversely, can be rejected as a whole, thus leading to the aggressive pruning alluded to above. Naturally, such families of design objects are incrementally discriminated, as we progress down in the specialization hierarchy.

For detailed examples of design discipline hierarchies, illustrating the various concepts discussed above, see [Jaco93][JaDi94][JaDi96].

A final note on terminology. As alluded to before, our early design space exploration framework considers the design space at the *system-level of abstraction* only. At this level of abstraction, *behavioral* and *structural* properties have been traditionally referred to as *algorithmic* and *architectural* properties, respectively. Thereby, from now on, behavioral/structural restrictions, descriptions, and requirements may alternatively be referred to as algorithmic/architectural restrictions, descriptions, and requirements, respectively.

2.3.4 Consistency Constraints

Properties may not be, and need not be *independent* from each other. In other words, properties that belong to the same, or different, abstraction levels, and to the same, or different, classes of design objects, may be constrained by arbitrarily complex relations. For example, the designer may make a design decision with respect to layout style that later on proves to be inconsistent with the performance and/or area re*quirements of the design object. Consistency constraints*, which represent such dependencies among properties, are defined in terms of an independent specification declaration, a dependent specification declaration, and a relation involving the properties contained in both specification declarations.

Consistency constraints should be defined so as to capture and enforce the design steps adopted in typical top-down design methodologies. Indeed this is particularly important since the design steps capture problem decomposition strategies used to obtain consistent (complete) design object specifications, subject to complex (sometimes circular) dependencies among the various object properties. Note that such strategies may lead to partial inconsistencies during certain phases of the design process, hence the need for validation steps.

Consider, for example, the *conceptual design* phase. Since the design decisions made during this phase may later on prove to be inconsistent with the system requirements, a number of consistency constraints should be defined, placing the relevant restrictions (i.e., design issues) on the independent specification declaration, and the corresponding requirements on the dependent specification declaration. Such consistency constraints thus state the need for verifying consistency between both sets of properties (i.e., define required verification steps).

Consider now the *transformational synthesis* steps typically undertaken during top-down design. Functional equivalence between the original behavioral descriptions and the synthesized (lower level) ones (say, between algorithm level and RTL behavioral descriptions) needs to be verified. Accordingly, a number of consistency constraints ought to be defined, each of which placing the object's behavioral description at a given level of abstraction (above the physical) on the independent specification declaration, and the immediately lower-level behavioral description on the dependent specification declaration. Such consistency constraints thus state that functional equivalence, between the more detailed and the less detailed descriptions, needs to be verified. They also define an ordering between the generation of such descriptions.

Consistency constraints can also be used to establish consistency relationships between design options for different design issues, and to define partial orderings between design issues.

Note finally that property definitions can be a member of any number of consistency constraints. Moreover, they may be listed as a member of the independent sub-specification for some of the consistency constraints, and as a member of the dependent sub-specification for the remaining consistency constraints -- the consistency constraints on functional equivalence alluded to above illustrate this last case.

This completes our brief overview of the design formalism used to characterize the design space in our early exploration framework. For an extensive discussion of the formalism and its applications see [Jaco93][JaDi96][LoJa92].

We are currently using the formalism briefly outlined above to derive a formal representation of the design space (at the system level of design abstraction) for a number of applications, including *encryption* and *video encoding/decoding*. (Note that our goal is not to prove that universal, complete design space characterizations can be created, but instead, to demonstrate that design space characterizations suiting the needs of particular classes of applications can indeed be created using our formalism.) Encryption poses particularly challenging problems, since it requires the abstract characterization of quite sophisticated algorithms, such as the Montgomery modular multiplication algorithm, whose optimized forms available in the literature [ElWa93] involve detailed (gate-level) implementation considerations, and may thus challenge abstraction principles. This formal representation of the design space will be incorporated in the framework's design space exploration sub-system, as shown in Figure 2.1.

2.4 The Algorithm and Architecture-level Models

As alluded to before, two models are maintained (within the design space exploration sub-system) to capture the design decisions made so far by the designer: (1) *algorithmic models* that represent the system behavior; and (2) *architectural models* that represent possible realizations of the system behavior. In this section, both models are reviewed in some detail.

2.4.1 Algorithm-Level Model

The algorithmic model is constructed by compiling the algorithmic description provided in the initial system specification into a set of hierarchical data flow graphs [PeJa97]. In such graphs, nodes represent computations/tasks and edges represent data dependencies [KaBu86][GaVa94].

The resulting model is hierarchical since *basic blocks*, defined by alternative (flow of) control paths, are first represented as single *complex nodes*. Such complex nodes are then recursively decomposed (i.e., expanded) into sub-graphs, containing less complex nodes, until the lowest level of granularity (i.e., the single operation level) is reached. These last nodes are designated *atomic nodes*. A complex node is thus a contraction of a sub-graph containing nodes of lower complexity (i.e., smaller granularity).

The graphs are polar, i.e., have the beginning and the end of their execution synchronized by a *source* and a *sink* nodes, respectively [Mich94]. Sink and source nodes have no computational cost, they are used simply to assert that sub-graphs contracted into complex nodes represent basic blocks defined by alternative control flow paths. For a detailed discussion on the various types of nodes supported in the algorithmic model and the *firing rules* defined for these node types see [PeJa97].

Complex nodes contain key information about their hierarchically dependent sub-graphs, including their critical path. Edges contain information on the data produced and consumed by the various nodes in the graph, including the specification of basic data type, encoding type, etc. Note that such information is essential for determining the specific physical resources required by each atomic node. (For example, some nodes may perform floating-point multiplications while other nodes may perform integer multiplications.)

In our proposed algorithmic model, hierarchy is represented by containment -- in order words, each graph may be considered at its higher level of abstraction or may have some of its complex nodes expanded into their corresponding sub-graphs. Thereby, the level of granularity of the model (and thus its size and complexity), can be dynamically adjusted, according to the needs of the design space exploration.

As mentioned previously, the algorithmic description provided in the initial system specification is required to be annotated with profiling information providing: (1) upper bounds and mean values for data dependent loop indices; and (2) relative frequencies of execution for basic blocks contained in alternative control flow paths (defined by "if-then-else" or "switch" conditional statements).

This information is important when, in the presence of data dependent constructs, one still wants to be able to reason about (average and worst case) execution delays and power consumption. Accordingly, based on the profiling information referred to above, a function is defined which associates a *transition probability* to every edge sourcing from a select node or sinking into a merge node. (Note that *select* nodes mark the beginning of alternative flow of control paths and *merge* nodes mark the end of such alternative control paths [PeJa97]) Such a function thus indicates the probability that the particular node will consume/produce data (i.e., tokens) from/to the specific edge. Obviously, the summation of the transition probabilities of all arcs sourcing (sinking) from a select (merge) node has to be one. This function is used to estimate the *average* and the *worst case* execution delays and power consumption for all nodes in the hierarchical model. The average and worst case execution delays, for example, are defined as the average path and longest path from source to sink, respectively.

As alluded to previously, the area, timing and power requirements (given in the system specification) are translated into corresponding *budgets* for the various nodes in the algorithmic model. These budgets establish upper-bounds on worst case and average execution delays, power consumption, etc., for each node in the hierarchy. They are defined top-down, considering the *relative* complexity of the various nodes in the model. During this budgeting process, hierarchy is exploited, in that the budgets of complex nodes are distributed through the nodes contained in their corresponding expansion graphs.

Feasibility is assessed by comparing the node budgets with the values *achieved* by the current node implementation. Node implementations are selected during architecture level design space exploration, which is the topic of the following sub-section.

So far, we have assumed that the requirements given in the initial system specification, and thus the corresponding budgets, specify average and *worst case* performance and power consumption. An important observation is that, sometimes, the application is amenable to some level of constraint relaxation. In such cases, some or all of the system requirements (e.g., on throughput and/or execution delay), and the corresponding budgets, are probabilistically specified. In [VeJa97] we have proposed an approximate algorithm that allows for quickly assessing probabilistic constraints on complex systems with uncertain delay components. The ability to perform such an analysis may lead to significant savings on the system's final cost (silicon area) and power consumption. The relevance and interest of constraint relaxation is discussed in [VeJa97] using the MPEG-II standard. In particular, MPEG-II decoders perform a number of tasks with data dependent (i.e., non-deterministic) delays, which, for high-performance implementations, can significantly impact their final cost. This last fact was demonstrated in [VeJa97] by considering the design of an MPEG-II decoder subject to probabilistic throughput requirements. The required profiling information (on distributions of picture and block types, etc.) has been derived using a number of video traces. For simplicity, in the remainder of the discussion we consider deterministic budgets only.

2.4.2 Architecture-level Model

As mentioned above, the design decisions made during design space exploration are compiled in an *architectural specification*, which includes: (1) a partitioning of the algorithmic model into a set of *algorithmic segments* (i.e., non-primitive behavioral building blocks), each of which is to be implemented by an individual *architectural component* (i.e., a non-primitive structural building block); and (2) a set of fundamental design decisions on the implementation of these architectural *components* and their *interfaces*.

Note that algorithmic segments can be implemented in *function-specific hardware components,* or can alternatively be implemented in software, meaning that the corresponding architectural component will be a *processor core*. In our approach, the decision to implement an algorithmic segment in hardware or in software is done by choosing an appropriate "implementation style" which eliminates library components which do not comply with the selected style.[2]

Before starting a detailed description of our proposed architecture-level model, an introductory discussion on the aim and scope of architecture-level design space exploration is in order. During such an exploration, the specific resources (or library components) that will execute each *atomic node* defined in the algorithmic model are selected from the algorithm and components library (considering the selected style). Those design decisions provide fundamental information on execution delay and power consumption for all of the nodes in the model. For example, a specific multiplier may be defined in the library (at the system-level of abstraction) as having an execution delay of 3 reference operations (normalized time measure) and as consuming 4 power units per reference operation (normalized power measure), while another resource, say, an adder, may only take 1 reference operation and consume a single power unit per reference operation.

Once each atomic node is mapped onto a physical resource, the hierarchy is traversed, bottom-up, and the worst case and average execution delay and power consumption for all complex nodes in the model are computed, based on the corresponding values for the atomic nodes.

Feasibility is then roughly assessed by comparing the node budgets with the actual values required by the current node implementations, which can be determined by resolving the reference operation. Nodes that violate their budgets, at any level of the hierarchy, can be thus quickly identified, and then *locally* optimized.

Note that the atomic nodes referred to above, implementing multiplications, additions, etc., correspond to *fundamental* operations in VHDL or C, the languages used to describe behavior at the system-level of abstraction. Accordingly, the functional units being selected to execute such atomic nodes correspond to *fundamental structural building blocks* at the system level of abstraction (i.e., adders, multipliers, etc.). Alternatively, complex nodes (i.e., nodes that contract sub-graphs of arbitrary complexity) may be directly mapped into complex modules, i.e., *non-fundamental structural building blocks*, if such modules are available in the library. In this last case, worst case and average *execution delay* and *power consumption* are directly provided in the library, and thus the corresponding complex nodes do not need to be further expanded.

So far we have mostly concentrated on defining libraries tailored to custom (i.e., function-specific) hardware designs. In these libraries, components of various granularities, such as multipliers, adders, and DCT modules, are characterized in terms of their delay, area, power consumption, voltage supply, etc. (Some of these measures/values are properly normalized, as referred to above.) Unless otherwise stated, the discussion from this point on focuses on function-specific hardware components.

---

[2] "Implementation style" is defined as a design issue.

The basic elements of the proposed architectural model are: *architectural components*, *modules*, and *physical resources*. In order to efficiently support architectural exploration, three additional elements were added to the algorithmic model: *algorithmic segments*, *clusters of nodes*, and *pipeline stages*.

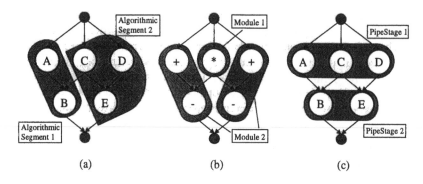

(a)   (b)   (c)

Figure 2.2. Example of (a) Algorithm Segments; (b) Clusters and Modules; and (3) Pipeline Stages

As referred to above, the original algorithmic model is partitioned into a set of algorithmic segments -- an *algorithmic segment* thus comprises one or more (sub)graphs belonging to an algorithmic model.[3] Figure 2.2(a), for example, shows two algorithmic segments defined on a simple algorithmic model that contains only one top-level graph with five basic block nodes. The implementation of an algorithmic segment (by an architectural component) is defined as the implementation of the various clusters of nodes defined for the particular segment, as discussed in the sequel.

A *cluster of nodes* is a set of one or more connected *atomic* nodes belonging to a graph within an algorithm segment. Figure 2.2(b), for example, shows three clusters defined within a particular graph. In order to completely define an architecture component, every atomic node within an algorithm segment must thus belong to one and only one cluster. Clusters of nodes are implemented by *modules*. A module consists of one or more physical resources (specifically, functional units), optimally interconnected for executing a specific cluster of nodes. Examples are modular exponentiation and modular multiplication modules, DCT and IDCT modules, multiply-accumulate modules, and adders. Note that clusters containing a single atomic node can be directly implemented by fundamental structural building blocks (i.e., adders, subtracters and multipliers), while more complex clusters typically require non-fundamental structural building blocks.

If no scheduling conflicts exist, modules can be shared among isomorphic clusters. For example, Module 1 in Figure 2.2(b) implements a multiplication and is

---

[3] Some transformations may need to be performed in nodes at the interface of a partition but, for simplicity, we omit them here.

used by only one single cluster, while Module 2 implements an addition followed by a subtraction and is shared by two isomorphic clusters.

We conclude the discussion of the function-specific architectural model by introducing the notion of *pipeline stage*. As mentioned in the previous section, the graphs that comprise an algorithmic description are polar, i.e., have a source and a sink node that synchronize the beginning and the end of the execution of the graph. By default, each of such graphs is fully enclosed in what we call a "pipeline stage" - - specifically, this means that the source node can only be fired after the sink node has completed its execution. This default scheduling policy can be modified, though, by increasing the number of pipeline stages defined within a graph. Figure 2.2(c), for example, shows a graph where two pipeline stages were defined -- the nodes enclosed in each of the two stages shown in the figure can now execute "concurrently" with respect to each other, thus increasing the throughput of the corresponding realization/implementation.

As alluded to above, most of our research effort in the area of architectural modeling (at the system-level of design abstraction) has so far been focused on function-specific hardware components. The research on system level models for software realizations has only recently started, and includes (1) deriving a systematic characterization of processor cores (with respect to basic architectural features, instruction-set, and clock frequency), aiming at establishing a basis for core selection during the design space exploration[4]; and (2) deriving instruction-level performance and power models for a number of processors, in order to support the rough feasibility analysis alluded to above.

2.5 Assessment of Candidate Solutions

As discussed previously, before dropping a less expensive architecture and move into a more costly one, it is important to make sure that "heuristically good" design points for the first architecture have been considered. As mentioned previously, a set of abstract metrics is provided to help designers to *quickly* converge to such desirable solutions. After one or more heuristically good design points are determined, hard numbers are *estimated*, in order to determine if the specified budgets are satisfied. In what follows we discuss our proposed abstract metrics and estimation mechanisms.

2.5.1 Metrics for Identifying Heuristically Good Solutions

One of the fundamental difficulties involved in accurately estimating performance, area, and power consumption at any level of design abstraction other than the physical level, is the ability to adequately account for the "physical resources" that are abstracted, i.e., not yet defined at the particular level.

---

[4] This characterization is being developed using the design formalism discussed previously, and will be added to the design space representation of the design exploration sub-system.

As suggested in [RaGu95][RaPe96], hardware resources can be broadly classified into two categories: algorithm inherent and implementation overhead. *Algorithm inherent resources* are the functional units (multipliers, adders, etc.) needed to implement the operations defined in the system's algorithmic description -- such resources are tangible and can be reasoned upon at the system-level of abstraction. *Implementation overhead resources* account for all of the other resources needed to correctly execute such operations, including control logic, steering logic (e.g., buses and MUXES), registers, and wiring.

A key observation is that implementation overhead resources are (typically) not yet defined at the system-level of abstraction -- recall that the idea is to assist designers in making fundamental design decisions concerning to the system architecture, before investing any significant effort in detailed design. An exception to this occurs when a complex node can be directly mapped on to a correspondingly complex module (say, an IDCT module, or a processor core). In this case, the implementation overhead internal to the particular module can be taken into consideration, since its detailed design is available in the library. However, in the extreme case, when there is no other choice but deciding on the system implementation by mapping *atomic nodes* on to basic functional units, (e.g., multipliers, adders, etc.), then implementation overhead is almost completely omitted from the rough feasibility analysis discussed in the previous section.

Unfortunately, implementation overhead can significantly contribute to the delay, power consumption and area of a design. Thereby, ignoring implementation overhead, by performing trade-offs only in terms of functional units, may lead to sub-optimal (i.e., inferior) design solutions. For example, at the structure level, one may attempt to trade-off speed for area by increasing the level of resource sharing in a given design. However, due to a possible increase in number and/or size of buses, MUXES, and registers, the overall area of the design may actually increase. Similarly, at the algorithmic level, one may attempt to reduce power consumption using, for example, *operation strength reduction* on power hungry operations, such as multiplication. A recent case study illustrates the results of applying one of such transformations -- specifically, "multiplications" substituted by "adds" and "shifts" -- to fourteen DCT algorithms, considering standard-cell implementations [PoKi97]. Interestingly, *in all cases* the overall power consumption has actually *increased* as a result of the transformation -- in half of the cases it actually increased by more than 40%. This was so because the power consumption in buses, registers and control circuitry has dramatically increased, in some cases by more than 400%, totally overshadowing the power savings in the functional units.[5]

In order to tackle this problem, we are working on a set of metrics designed to rank competing solutions based on their potential to take advantage of specific

---

[5] Note also that the impact of implementation overhead on the performance and power consumption of processor cores is typically more significant than it is for custom modules (due to the more "flexible" nature of those programmable components). Specifically, the execution delay of an instruction, as well as its power consumption, are typically much larger than those of the involved *functional* unit(s), hence the interest of the instruction level performance and power characterizations alluded to above.

attributes/properties that correlate to *minimal implementation overhead*. Examples of such attributes are given below.

2.5.1.1 Locality of Computations

A group of computations within an algorithm is said to have a high degree of *locality* if the algorithm level representation of those computations corresponds to an isolated, strongly connected (sub)graph [ScSt93]. More informally, the largest the volume of data being transferred among the nodes belonging to the "cluster", in comparison with the volume of data entering and exiting the "cluster", the highest will be the degree of locality. By considering such strongly connected sub-graphs as single clusters of nodes, and thus implementing them using modules optimized for performing the specific set of computations and data transfers, it is possible to *minimize* the implementation overhead needed to execute the corresponding part of the algorithm. Namely, the required physical resources will be located in the proximity of each other, thus minimizing the length of interconnect and/or buses. So, if an algorithm exhibits a good degree of locality of computations, i.e., has a number of isolated, strongly connected clusters of computation, such clusters define a way of *organizing the functional units into modules*, so as minimize the number of global buses, and possibly other implementation overhead resources. Accordingly, solutions that adhere to, or favor, such an organization are considered to be "heuristically good." Moreover, if the relevant budgets allow for a software implementation of the strongly connected cluster, then the entire cluster should be mapped on to a single processor.

2.5.1.2 Regularity

If a given algorithm, or algorithmic segment, exhibits a high degree of *regularity*, i.e., requires the repeated computation of certain patterns of operations, it is possible to take advantage of such regularity for minimizing implementation overhead. Modules can thus be optimized for executing not only one, but a number of isomorphic clusters, thus leading to solutions that minimize the "cost" of resource sharing in terms of implementation overhead. We measure the degree of regularity of an algorithm by the number of isomorphic sub-graphs that can be identified in its corresponding algorithmic model. As above, solutions that exhibit a high degree of regularity are considered to be "heuristically good."

2.5.1.3 Sharing Distance

This metric considers the "distance" between clusters of nodes that share the same module. The idea is to favor solutions that maintain a certain degree of locality in their module sharing policy, thus minimizing the need for global buses. Note that the "distances" referred to above are measured in "time", i.e., in number of reference operations between two consecutive executions of the same module. Accordingly, this metric can also be used to favors solutions that concentrate the use of modules in specific time intervals, thus creating opportunities for *clock gating* [RaPe96].

The above measures should be carefully considered during the design space exploration, as briefly discussed in the sequel. Locality of computations and regularity, for example, are intrinsic behavioral properties -- they can thus be used to initially drive the algorithmic exploration process. Taking maximum advantage of these attributes may require an aggressive architecture- level design space exploration, though. For example, an obvious goal is to fully implement strongly connected clusters of computation within a single architectural component, be it a function specific hardware component (possibly comprised of several modules), or an embedded processor core. Well known partitioning algorithms are available in the literature, aiming at minimizing interconnections among architectural components [AlHu97][KaAg97]. A less well studied problem, though, is to attempt to define algorithmic segments (i.e., partitions) that maximize the degree of regularity within function-specific components. Since this strategy may lead to significant savings in implementation overhead, it clearly illustrates the benefits of an aggressive, early design space exploration. Finally, within a given component, trade-offs between larger modules, optimizing large chunks of computations, vs. smaller modules, shared by a number of smaller isomorphic clusters, should also be carefully considered.

Naturally, the system's timing, power, and area budgets must be taken into account throughout the entire design space exploration, since they define the requirements that must be met by the implementation. For example, if a given computationally expensive loop has stringent timing constraints, its parallelization may need to be considered, in an attempt to increase execution speed, even if locality will most probably be compromised by such a parallelization.

2.5.2 Assisting Floorplanning and Early Performance Estimation

Today it is possible to create transistors with widths below 0.25 microns, for what is referred to as deep submicron CMOS technologies [Wolf95]. These submicron feature sizes yield transistors that operate at higher speeds and permit more transistors to be placed on a single integrated circuit. However, the advances toward dense circuits have resulted in the digital IC area no longer being limited by the transistors that are placed on the silicon, but by the metal interconnect which must connect them. Furthermore, the interconnect is becoming itself an increasingly significant contributor to on-chip signal delay. In short, *modularity* and *locality*, the fundamental principles enabling the use of abstraction during hierarchical decomposition, will sooner no longer hold.

In order to address this problem, we are working on a methodology for assisting floorplanning in the context of implementation technologies for which modularity and locality do not hold. The objective is to be able to still generate accurate performance estimations under such conditions, and thus support hierarchical, top-down designs. We propose to generate such early predictions based on statistically driven inferences from previous designs. Specifically, in order to assess the performance of "new" designs where no layout is available, we propose an innovative approach based on the creation of virtual layouts, i.e., modular compositions of existing layouts, which in turn are used to generate performance predictions for the eventual physical designs.

We are currently developing mechanisms for: (1) generating such virtual layouts; (2) assessing the quality of the approximation provided by a virtual layout; and (3) generating performance estimates that reflect the confidence on the quality of the approximation.

One of the critical issues being currently investigated is thus the ability to evaluate the quality of the approximation provided by a virtual layout, and then map that into a degree of confidence in the estimates generated using the approximation. Statistical quantifiers of quality, targeting layout densities, and distributions of net lengths, capacitances and resistances, are being investigated for the purpose. Such quantifiers are being derived using the layout samples available in our library of algorithms and components.

2.6 Defining Libraries of Algorithms and Components

This section discusses the library of algorithms and components shown in Figure 2.1, which defines the universe of algorithmic and architectural alternatives viable for implementing the system designs.

As discussed before, we are currently experimentally validating our approach using a number of applications, including encryption and video encoding/decoding. The library under development will thus contain from simple to highly optimized, complex design objects, such as modular exponentiation and modular multiplication modules, DCTs, IDCTs, adders, etc., developed using various algorithms and implementation styles.

A key observation is that all of the key concepts and principles used to characterize the design space (in the design-space exploration sub-system) are mirrored in the library, as discussed in the sequel. Specifically, the myriad of design data characterizing the design objects stored in the library is "discretized" in terms of a number of design data segments, each of which is "indexed" by a property. Design abstraction is also explicitly represented in the library, so that different abstraction hierarchies can be created and co-exist, suiting the needs of specific families of implementation styles (such as application-specific hardware modules, or software executing on embedded cores). Specifically, the properties describing each design object resident in the library are partitioned into adequate levels of abstraction, forming the design object's facets. The concept of specialization is also implemented within each level of design abstraction, in order to supporting the effective design space pruning alluded to previously.

Let us illustrate how the set of alternative IDCT designs currently under development, for example, will be organized in the library of algorithms and components. At the system-level of design abstraction, a sub-hierarchy defining the *IDCT "family" of design objects* is defined. The "root" of this sub-hierarchy specifies a generalized "system-level" facet , or view, of the *IDCT class of design objects* -- such a generalized facet has very little information on it, beyond the definition of the discrete cosine transform itself [IfJe93]. Progressing down in the specialization hierarchy, a number of "families" of algorithms for computing discrete cosine transforms are then specified -- such "families" are to be constructed based on algorithm similarities (such as commonalties on the number of operations on the critical path, precision, etc.) that are *relevant* to performance, power consumption, and other typical design requirements. Progressing further down in the specialization hierarchy, alternative implementation styles for these algorithms start to be discriminated -- those may include application-specific datapaths, software implementations running on an embedded core, or a mix of both. In short, families of algorithms and realizations, materializing different cost/power/ performance *ranges* and *trade-offs*, are first defined. Such "families" can be thus directly inspected and selected/discarded by the designer, thus allowing for a systematic pruning of the design space, from inexpensive to increasingly costly alternative IDCT designs.

Naturally, as new designs are produced in the design environment, they can be added to the library, thus providing new design points to be used in further sessions.

In spite of the fact that the framework being discussed is geared at supporting system-level design space exploration, the designs stored in the library can be defined down to the of physical level design abstraction, whenever such a level of detail is available. As discussed in the previous section, these lower-level facets can be used to generate ("on-the-fly") the layout approximations needed for assessing performance and other physical characteristics of the IC under design.

2.7 Concluding Remarks

We have presented a methodology for early system level design space exploration. The goal is to assist designers in systematically identifying cost-effective system-on-chip architectures for various classes of applications, including consumer electronics and communication products.

The size, complexity, and heterogeneity of such designs poses a number of challenging research problems, including: (1) the need for a formal characterization of the design space; (2) the need for effective mechanisms to comparatively assess the quality of alternative solutions and to perform a crude feasibility analysis on these solutions; (3) the need for mechanisms for early performance estimation for deep-submicron designs; and (4) the need for design libraries that can effectively support the early, system-level design space exploration.

Preliminary solutions to these problems have been proposed and discussed in the paper. The development of the framework for system-level design space exploration, outlined in Figure 2.1, is ongoing.

Acknowledgments

Besides the support of the EC, this work has been also funded by the National Science Foundation (MIP-9624321, Jacome), and by the NATO (CRG 950809, Jacome and López).

References

[AlHu97] C. Alpert, J. Huang and A. Kahng, "Multilevel Circuit Partitioning", in Proceedings of *34th ACM/IEEE Design Automation Conference*, Jun 1997, pp. 530-533.

[ElWa93] S. Eldridge and C. Walter, *"Hardware Implementation of Montgomery's Modular Multiplication Algorithm"*, in IEEE Transactions on Computers, Vol. 42, No. 2, Jun 1993, pp. 693-699.

[GaVa94] D. Gajski, F. Vahid, S. Narayan and J. Gong. *Specification and Design of Embedded Systems*. P T R Prentice Hall, 1994.

[Goos96] G. Goossens et al, "Programmable Chips in Consumer Electronics and Telecommunications", chapter in *Hardware/Software Co-design*, G. DeMicheli and M. Sami, eds., NATO ASI Series Vol. 310, Kluwer Academic Publishers, 1996.

[Gren66] U. Grenander, *Elements of Pattern Analysis*, Johns Hopkins University Press, 1966.

[IfJe93] E. Ifeachor and B. Jervis, *Digital Signal Processing, A Practical Approach*, Addison-Wesley Publishers Ltd., 1993.

[Jaco93] M. Jacome, *"Design Process Planning and Management for CAD Frameworks"*, Ph.D. thesis, Carnegie Mellon University, Department of Electrical and Computer Engineering, Sep 1993.

[JaDi92] M. Jacome and S. Director. "Design Process Management for CAD Frameworks." In Proceedings of *29th ACM/IEEE Design Automation Conference*, Jun 1992, pp. 500-505.

[JaDi94] M. Jacome and S. Director. "A Formal Basis for Design Process Planning and Management", in Proceedings of *ACM/IEEE International Conference on CAD*, Nov 1994, pp. 516-521.

[JaDi96] M. Jacome and S. Director, *"A Formal Basis for Design Process Planning and Management"*, in IEEE Transactions on Computer-Aided Design of Integrated Circuits and Systems, Vol. 15, No. 10, Oct 1996, pp. 1197-1211.

[KaAg97] G. Karypis, R. Aggarwal and S. Shekhar, "Multilevel Hypergraph Partitioning: Application in VLSI Domain", in Proceedings of *34th ACM/IEEE Design Automation Conference*, Jun 1997, pp. 526-529.

[KaBu86] K. Kavi, B. Buckles and U. N. Bhat, *"A Formal Definition of Data Flow Graph Models"*, IEEE Transactions on Computers, C-35 (11), 1986.

[LoJa92] J.C. Lopez, M. Jacome and S. Director, "Design Assistance for CAD Frameworks", in Proceedings of *GI/ACM/IEEE/IFIP European Design Automation Conference*, Sep 1992, pp. 494-499.

[Mich94] G. DeMicheli. *Synthesis and Optimization of Digital Circuits*. McGraw-Hill, 1994.

[MiSa96] G. DeMicheli and M. Sami (eds.), *Hardware/Software Codesign*, NATO ASI Series Vol. 310, Kluwer Academic Publishers, 1996.

[PeJa97] H. Peixoto and M. Jacome, "Algorithm and Architecture Level Design Space Exploration Using Hierarchical Data Flows", In Proceedings of *11th International Conference on Application-specific Systems, Architectures and Processors*, Jul 1997, pp. 272-282.

[PoKi97] M. Potkonjak, K. Kim and R. Karri, "Methodology for Behavioral Synthesis-based Algorithm-level Design Space Exploration: DCT Case Study", in Proceedings of *34th Design Automation Conference*, Jun 1997, pp. 252-257.

[RaGu95] J. Rabaey, L. Guerra and R. Mehra, "Design Guidance in the Power Dimension", In Proceedings of *ICASSP 95*, May 1995.

[RaPe96] J. Rabaey and M. Pedram (eds.), *Low Power Design Methodologies*, Kluwer Academic Publishers, 1996.

[ScSt93] G. Schmidt, T. Strohlein, *Relations and Graphs - Discrete Mathematics for Computer Scientists*, Springer-Verlag, 1993.

[Smal96] C. Small, *The Statistical Theory of Shape"*, Springer-Verlag, 1996.

[VeJa97] G. Veciana and M. Jacome, *"Hierarchical Algorithms for Assessing Probabilistic Constraints on System Performance"*, Technical Report, The University of Texas at Austin, Mar 1997. (also submitted to a conference)

[Wolf95] S. Wolf, *"Silicon Processing for the VLSI Era. Volume 3 - The Submicron MOSFET"*, Lattice Press, 1995.

3 KNOWLEDGE BASED HARDWARE-SOFTWARE PARTITIONING OF ELECTRONIC SYSTEMS

María Luisa López-Vallejo
Juan Carlos López

3.1 Introduction

The growing complexity of today's electronic systems implies that many designs have either too many components to fit in a single chip, or strong marketing and technical constraints making unfeasible the implementation of the whole system on a single dedicated circuit. In addition, electronic designs are driven more and more by shrinking market windows that force the designers to come up with new methods to deal with such puzzling designs. The goal of shortening the *time-to-market* of a specific product, while increasing design quality, has been traditionally achieved by moving up the abstraction level in which the design is specified. Thus, using the proper tools, the designers can get rid of implementation details while focusing on a better search through larger design spaces.

The ability of using hardware and software components that can be simultaneously designed appears as an alternative to achieve such goals. In fact, complex systems with components of a very different nature are the protagonists of the electronic design scene. *Hardware-software co-design* tackles these issues by providing methodologies and tools for the fast development of complex heterogeneous systems.

In the search for good system implementations many design tradeoffs have to be examined. Decisions made at very high level, that is, in the early phases of the design process, clearly make an impact on all the subsequent stages. Among these

J.C. López et al. (eds.), Advanced Techniques for Embedded Systems Design and Test, 53-76.
© 1998 Kluwer Academic Publishers.

decisions, the selection of components that should be implemented either in the hardware or the software is of crucial importance, and, therefore, *hardware-software partitioning*, as the design stage where this issue is faced, turns out to be a key task in the system design process. This chapter will rationalize the importance of this phase and will present a different way of addressing the hardware-software partitioning problem.

3.2 The Co-design Problem

The main objective of the design automation community is to provide methods and tools to design high performance systems at the lowest price, two antagonistic goals that must come together in a common approach. To reach such an objective, it is necessary to look for the combination of the system components that achieves the best implementation. *Hardware-Software Co-design* deals with this problem by combining the hardware and software development cycles and searching for the best tradeoff among the different solutions. The basic idea is to delay the selection of the most suitable implementation for every system block as much as possible, trying to automate all the tasks involved in the process.

3.2.1 Motivation: Design of Heterogeneous Systems

The current trend is for digital designs to have many programmability requirements that demand the use of programmable components (standard microprocessors, microcontrollers, etc.) allowing a more flexible design. *Heterogeneous systems* (those with both types of components: programmable and dedicated) are used more and more: products for mobile communications, consumer electronics, etc. are, de facto, heterogeneous systems with hardware and software parts.

The necessity of dealing with programmable elements has caused an essential modification of the classical design process so as to integrate in it the development phase of such programmable components. This is the *Hardware-Software Co-design* responsibility: the generation of tools for this special design cycle that must produce good results in an appropriate time.

To sum up, the main advantages of following a co-design methodology for the design of heterogeneous systems are:

- The global latency of the system is, in general, decreased, since hardware and software parts will execute concurrently.

- The use of a hardware co-processor allows execution speedup for critical modules.

- Flexibility is guaranteed by the use of programmable components.

- The tradeoff in flexibility, good performance and cost can be better found.

- Finally, the design cycle is reduced, resulting in a shorter development process. It is important to remark that the main objective of a product design process is to minimize as much as possible its *time-to-market*.

3.2.2 A typical Co-design Methodology

Even though every co-design environment or methodology has its own singularities, there are several basic processes or tasks that can be found in all of them. The scheme in Figure 3.1 shows the way these processes interact. These tasks are:

- System specification using high level languages.

- Formal transformations and optimizations.

- Translation of the initial description into an intermediate format, which eases the handling of information in later steps. This is the base of the system model for partitioning.

- Hardware-software partitioning (performed automatically or manually).

- Synthesis of the hardware blocks.

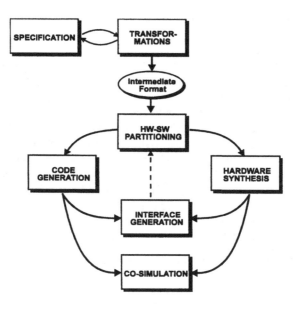

Figure 3.1. Design flow of a typical Co-design Methodology

- Machine code generation for the software blocks.

- Interface generation (integration of hardware and software parts).

- Verification of the system functionality and the satisfaction of the system constraints (very often done by co-simulation).

3.3 Partitioning as a Key Issue in Co-design

3.3.1 The Role of the Partitioning Phase

As discussed before, it can be stated that hardware-software partitioning is one of the main phases during the design of complex systems. During this process it is established which parts of the system specification will be implemented as software running in a standard processor and which parts will have to be implemented in special purpose hardware. But the partitioning procedure is not just an assignment of blocks of functionality to hardware or software implementation units. It must carry out three main tasks:

1. The assignment of fragments of the original system description to the different kinds of system components: ASICs, standard microprocessors or micro-controllers, memories, etc.

1. The proper scheduling of those blocks of specification into time slots.

2. The allocation of the assigned and scheduled blocks to particular implementation units.

Partitioning is a key process in co-design since the decisions made at this time will directly affect the performance and cost of the final implementation. Therefore, the main objective of partitioning is to find a solution that fulfills the description functionality with minimum cost (or acceptable cost if the optimum solution is not reachable) and satisfying the global system goals and constraints. To succeed in such a complex objective many algorithms and methods are currently under development and research. Commonly, global optimization methods have been taken as foundation for these algorithms. Section 3.4 will review these methods, while a new approach based on artificial intelligence techniques will be presented in section 3.5.

3.3.2 Characterization of the Partitioning Problem

Hardware-software partitioning can be carried out by means of different techniques and methods, like adaptation of classical partitioning algorithms (min-cut [KL70], clustering [Lag89], etc.), general optimization methods (simulated annealing [KGV83], tabu search [GTW93] etc.) or newer approaches like expert systems or the application of genetic algorithms. In spite of the great variety of algorithms some common points can be observed:

- All of them are general optimization procedures guided by a global cost function.

- There is always structural support based on graphs.

- Some specific metrics are always needed: hardware area and performance estimates, software execution time and capacity (memory size for program and data), and some measure of the communication overhead in the hardware-software interface.

In this section some parameters will be defined in order to characterize the partitioning procedures and consequently evaluate the quality and properties of the different partitioning techniques. Since a general partitioning strategy can be viewed as *an algorithm* that assigns objects of a *system model* to *architectural components* guided by a *cost function* within a given *application domain*, four parameters have been chosen for this purpose: the system model, the cost function that guides the process, the specific architecture selected to implement the system and the type of application. They will be explained in detail in the next sections.

3.3.2.1 System Model for Partitioning

As it has been previously pointed out, before starting the partitioning process the initial system description is translated into an intermediate representation better suited for later manipulation. This representation is the mapping of the system model and this does not only hold a data structure but also the behavior and the information required by the partitioning procedure. This model does not necessarily correspond to the specification model, since it should be more adequate for partitioning issues and algorithms. Well known models currently used are:

- Control/Dataflow graphs,

- FSM based representations,

- Petri nets, and

- Structural graphs.

It can be seen that these intermediate formats are based on graphs whose vertices are the basic objects handled during partitioning. Vertices can stand for very different concepts, according to the abstraction level, component size, etc. Therefore, the internal model defines the partitioning objects. Relating to the object size, a specific parameter can be defined: the *granularity*. Given a system description, the granularity determines how to decompose the initial specification into a set of objects which will be distributed into the various system components. There are two basic types of granularity:

- **Coarse granularity:** the partitioning objects are major fragments of the initial specification. Typical examples are processes, tasks or subroutines. Some advantages of this granularity are smaller design space exploration, and consequently reduced computation time and easier designer interaction.

- **Fine granularity:** the partitioning objects are basic operators or instructions. This kind of granularity produces very large search spaces and consequently the calculation time grows exponentially. Furthermore, the results become difficult to understand, but eventually the obtained solution can be better, since a larger space of possible solutions is considered.

Although most partitioning approaches have a fixed granularity (coarse or fine) a better approach should consider the combination of objects with different granularity, what would provide higher versatility. A great effort is being made in this new area of research [HE97].

3.3.2.2 Cost Function

The evaluation of the suitability of the implementation of partitioning objects in hardware or software is not straight-forward. Many factors can be combined to conclude this preference. The way this evaluation is performed is computing the quality of a given partition by means of a *cost function* which draws on factors like:

- *Fabrication costs,* basically hardware area and required memory size. Recent efforts concentrate on low power design.

- *Performance related aspects,* like response times, throughputs, global latency or working frequency.

- *Constraint violation,* usually maximum area for the special purpose hardware and global timing conditions.

Defining system level cost functions is not easy, because different and sometimes opposing metrics must be considered. Hardware-software developers assume, as a general rule, that a hardware implementation requires high costs and a long design cycle, but provides the best performance. On the other hand, a software implementation is more flexible, cheaper and with a short design time, but supplies lower performance. Based on these assumptions most partitioning algorithms try to meet timing restrictions implementing time consuming blocks in hardware while keeping the global area constraint safe.

All these costs that guide the partitioning procedure are normally based on estimates. This is due to the long time needed to compute exact values along the co-design cycle (high level synthesis, code generation, compilation, etc.) for every possible configuration.

3.3.2.3 Target Architecture

The *target architecture* is a description of the number and type of components available to build the system and the way these components interact by means of special kinds of connectors. Co-design architectures are characterized by including at least a programmable component (a standard processor, a microcontroller, etc.) and a dedicated hardware device (currently an ASIC). Thus, the simplest co-design architecture is composed of a standard microprocessor and an ASIC using as a communication mechanism a shared memory accessed via an interface bus. It is not possible to classify the system architectures found in the literature due to the very different application domains and purposes. Nevertheless, there are four main architectural tendencies that can be enumerated:

1. Classical architectures based on the scheme: standard microprocessor + ASIC + memory + interface bus [GM93, EHB93, CLL96, EPK97]

1. A data oriented version of the classical architecture which is based on the use of DSPs [KL94, BML97]

2. System-on-a-chip architectures, which use embedded core processors within application specific circuits [COB95, CGJ94].

3. Real time distributed systems [Wo97].

It is clear that the architecture chosen to implement a given complex system drastically biases the partitioning procedure. Consequently, most co-design environments set up a fixed architecture and develop all the tools with this parameter already established. Nowadays, a great deal of time is spent looking at how to select the architecture of the system by quickly exploring all possible solutions (the reader is referred to Chapter 2 of this book). Although this would increase the development cycle (an additional phase is introduced), better results can eventually be obtained according to the particular problem dealt with.

3.3.2.4 Application Domain

It is unreal to think about building a general purpose co-design environment, because it is impracticable. The fact is that all the hardware-software partitioners are application driven. The main classification that can be made is based on distinguishing between *data oriented* and *control oriented* systems. The partitioning method must be different in both cases, since important issues for data processing systems probably will not matter in control based systems. For instance, the granularity used to perform the partitioning can be different. On the one hand, control oriented systems are normally based on coarse granularity because they usually have hard input-output restrictions. On the other hand, fine granularity is more suitable for data processing systems because they are composed of big datapaths with hard and low level latency constraints on them.

3.4 Previous Related Work

To perform hardware-software partitioning a great number of algorithms have been developed in different co-design frameworks. As it has been previously remarked, it is impossible to make a comparative study of partitioning methods since the premises and initial conditions of the varied environments strongly differ. Nevertheless, a taxonomy of co-design partitioning approaches can be proposed based on the characteristics explained in section 3.3.2. Table 3.1 shows the results of this rough classification (just the most well-known co-design system have been considered).

Procedure	*Granularity*	*Algorithm*	*Group*
Automated	**Coarse**	List scheduling based	Ptolemy (U.C. Berkeley)
		Hardware partitioning	F. Vahid (U.C. Riverside)
		Heuristics (SA, TS)	P. Eles (Univ. Linköping)
	Fine	Simulated Annealing	Cosyma (Univ. Braunschweig)
		Group migration	Vulcan (Stanford Univ.)
		Dynamic programming	Lycos (Tech. Univ. Denmark)
Manual	Coarse		Polis, Chinook, CoWare

Table 3.1. Taxonomy of hardware-software partitioning procedures

3.4.1 Automated Approaches

3.4.1.1 Coarse Granularity

The **Ptolemy Environment** [KL94, KL97] is focused on embedded real time systems for signal processing. The input to the partitioning tool is a directed acyclic graph of system tasks. Consequently, the granularity of the algorithm is *coarse*. Graph nodes are annotated with tasks information obtained by estimation. Given a system specification divided into fragments (tasks), partitioning must minimize the hardware area while meeting timing constraints, and it can be done in two ways:

1. The *Binary Partitioning Problem* [KL94], consisting of (1) the assignment of hardware and software implementation units for every task, and (2) the global system schedule. It is performed by the *GCLP* (Global Criticality and Local Phase) heuristic following a list scheduling strategy.

2. The *Extended Partitioning Problem* [KL97], which extends the binary partitioning performing a third procedure: the selection of the best architecture for the hardware modules. This problem is solved by the *MIBS* (Mapping and Implementation Bin Selection) algorithm, that is based on the use of a

lookahead measure to correlate the implementation bin of a given node with the hardware area required for the nodes not yet assigned.

The *cost function* that guides the partitioning process is composed of two contradictory objectives that are interchanged according to the particular algorithm state. These two objectives are: (1) to reduce the final execution time of a task, or (2) to minimize resources (hardware area or memory size and communication implementation).

The **GPP** (General Purpose Partitioner) [VL96,VL97] is a system partitioning platform developed by F. Vahid et al. This group has taken advantage of prior work on hardware partitioning, and has extended several circuit partitioning algorithms to perform hardware-software partitioning (the Kernighan and Lin heuristic, hierarchical clustering, simulated annealing, etc.). The main characteristics of this approach are:

- A classical target architecture,

- High level specification language (VHDL or a VHDL extension, SpecCharts),

- A particular system model: a directed acyclic access graph (the SLIF Graph [VG95]).

- Estimation as the basic procedure used to generate the measures that guide the different cost functions.

P. Eles [EPK97] proposes also modifications to classical optimization algorithms: Simulated Annealing [KGV83] and Tabu Search [GTW93]. This approach is strongly based on the specification language, a VHDL extension which includes a synchronous message passing mechanism for processes communication. From this specification a *coarse grain* intermediate representation is constructed. The basic objects considered during partitioning are subprograms, loops or blocks of statements, which are selected by identification of performance critical regions. The *target architecture* is composed of a standard microprocessor executing the software processes (with a run-time system performing dynamic scheduling) and a hardware coprocessor working in parallel.

The *cost function* drives the algorithms to achieve maximum performance while minimizing communication costs between the software and hardware partitions and improving the overall parallelism. The metrics used to evaluate the partitioning objects are obtained from profiling, static analysis and cost estimation.

3.4.1.2 Fine Granulariry

COSYMA [EHB93], one of the first co-design environments that appeared in the early nineties, implements a "software oriented" approach, because the starting point is an all-software solution from which blocks are extracted to hardware until system constraints are met (basically execution time and hardware area). The *target architecture* is the classical system composed of a standard SPARC processor that communicates with a synthesized coprocessor via shared memory. The processor and the coprocessor cannot work concurrently. The *algorithm* chosen to implement partitioning in this way is simulated annealing [KGV83], which extracts time consuming basic blocks from the input graph (an extended syntax graph) to be implemented in hardware. Therefore, the approach has fine granularity.

To reduce the computation time, partitioning is performed in two stages:

1. The "Outer partitioning loop", where time consuming computations are performed (hardware synthesis, software compilation or time analysis). It is executed very few times.

2. The "Inner partitioning loop", whose running time is smaller and consequently can be iterated more often (in fact, simulated annealing is performed here). Estimates are used in this case.

The simulated annealing is guided by a *cost function* based on two parameters: the potential performance improvement (speedup) and the communication overhead. Its evaluation is performed incrementally, to reduce the execution time per movement.

Gupta and de Micheli proposed in **Vulcan** [GM93] a *"hardware-oriented"* approach: it starts with an all hardware solution and, to reduce costs, the design system gradually moves blocks from hardware to software while constraints are not violated. The initial system description (specified by means of a hardware description language) is translated into a system model based on a set of flow graphs (CDFG). Communication within the flow graphs is performed via a shared memory while communication across different flow graphs is based on the message passing paradigm.

The algorithm has a fine granularity, because the basic partitioning units are language-level operations. Partitioning is performed by *heuristic search* (a group migration algorithm) guided by a *cost function* that considers the following assumptions:

1. Timing constraints are satisfied,

2. The processor and bus utilization are within requirements, and

3. The cost function is minimized.

This cost function considers hardware area, memory size and processor and bus utilization.

The **LYCOS** [MGK97] co-synthesis system starts with C or VHDL input specifications and looks for the hardware-software partition that minimizes the execution time for a given hardware area restriction. The *target architecture* is composed of a single microprocessor and a single dedicated hardware element connected via a communication channel (communication is made by means of synchronous transfers through a shared memory). The approach has fine granularity because the partitioning objects are the description basic blocks (BSBs). Hardware and software BSBs cannot execute concurrently The partitioning *algorithm* (called PACE) is based on the dynamic programming algorithm SimpleKnapsack. It uses BSBs sequences instead of single BSBs to reduce the algorithm complexity.

3.4.2 Systems with Manual Partitioning

Many co-design environments do not perform automated hardware-software partitioning. Instead, these systems provide a useful framework to help the designer when assigning high level modules to hardware or software. Therefore, design exploration is performed by the designer within an environment that support the decision variables: estimation, co-simulation, etc. This is the case of the following environments: Chinook [COB95], Polis [CGJ94] or CoWare [BML97].

3.5 A Fuzzy Logic based Expert System for Hardware-Software Partitioning

In this section a knowledge based system that performs hardware-software partitioning is described. SHAPES (*Software-Hardware Partitioning Expert System*) is a fuzzy-logic based expert system which provides an easy way to address hardware-software partitioning. The main benefits of the use of this kind of tools are:

- The possibility of dealing with imprecise and usually uncertain values (by means of the definition of fuzzy magnitudes).

- The application of the designer knowledge in the decision making process.

These techniques have been recently successfully applied to different and important design automation phases, like analog circuit synthesis [TCF96] or test space exploration [FK94].

3.5.1 Why Artificial Intelligence Techniques for Hardware-Software Partitioning

In the traditional design style, heterogeneous system development was fully characterized by an initial partitioning phase which drastically influenced the rest of the design steps. In this phase, a system designer decided which blocks of the system could be implemented in hardware and which could be realized in software, taking into account different pieces of estimation and basically his/her knowledge as an expert in the field. To automate this partitioning phase, it is necessary to mimic the way a skilled designer performs this step. As it has been shown in section 3.4, currently most approaches use complex algorithms to solve the problem, introducing complicated cost functions with the purpose of combining different and opposed objectives.

But the most suited way to implement a tool that "*simulates*" the designer behavior is the use of an **expert system**. Expert systems place the experience of specialists into the memory of a computer which afterwards works with it. The use of an expert system for hardware-software partitioning has many advantages:

- Expert systems emulate the way human experts think.

- Non experienced users can have benefit of another's knowledge acquired by experience.

- The process can be traced.

- The knowledge of the system can be increased as long as the system is used with different cases (learning is open).

- They do not only work with data but also with knowledge (including the subjective knowledge implicit on the linguistic information provided by the expert)

The best way to express the subjective knowledge implicit in the expert assessments is using **fuzzy logic** [Za73]. Appendix A introduces some basic definitions in fuzzy set theory. Sentences like:

"*if the hardware area of a block is not too large, and its execution time is considerable, it is suitable to be implemented in hardware*"

can be easily reproduced by a fuzzy conditional statement in the form 'IF A THEN B'. In this example three linguistic variables can be defined: *hardware area*, *execution time* and *implementation*. These variables take respectively the values *not too large*, *considerable* and *hardware*. Their association postulates a rule.

In the development of SHAPES several rules for the design of complex systems have been collected and put together to implement the knowledge bases of a partitioning expert system. Other important arguments corroborate the use of fuzzy logic, like:

- The knowledge stored in the base of rules of the expert system has been provided by humans, being therefore imprecise in nature. Crisp values are not suitable to represent this kind of information.

- There is also uncertainty related to the knowledge. The inference engine of the system must be able to analyze the transmission of uncertainty from the antecedents to the consequents of the rules.

- Its simplicity and proved effectiveness [Me95].

3.5.2 The SHAPES System Model

The input to the partitioning process is an execution flow graph which comes from the initial system specification. This is a directed and acyclic graph where nodes stand for tasks and edges represent data and control dependencies, following a coarse grain approach. There are several reasons to choose coarse granularity [LCL96]:

- System designers do not work with operations or instructions. Thus, fine granularity does not allow to model the knowledge in an expert system.

- The initial specification is a high level description, therefore it is reasonable to maintain the processes defined by the designer as partitioning objects.

- Designer interaction is allowed and therefore manual changes are supported.

- The complexity of the problem is smaller because the search space is reduced.

- Resource sharing can be better dealt with.

Every graph node is labeled with additional information about the processes. In detail, these pieces of information for a node i are: hardware area (ha_i), hardware execution time (ht_i), software execution time (st_i) and the average number of times the task is executed (n_i). Edges also have a weight associated $(comm_{ij})$ related to the communications between nodes i and j. This estimate is obtained from three different values: the transfer time $(t_{trans(i,j)})$, the synchronization time $(t_{synch(i,j)})$ and the average number of times the communication takes place $(n_{i,j})$.

As it has been pointed out, system partitioning is clearly influenced by the target architecture onto which the hardware and software will be mapped. Here the target architecture considered follows the classical scheme of a processor running the software, an ASIC implementing the hardware and a shared memory accessed through a common bus. Interface modules are used to connect the processor and the ASIC to the bus.

The output of the partitioning tool is not only an assignment of blocks to hardware or software implementations, but also their scheduling (starting and finishing time) and the communication overhead produced in the interface. Collateral tools are required to build the system model that serves as input of the partitioning expert system. First a parsing stage from the initial specification extracts the data structure. A profiler is used to evaluate the software performance (st_i), while high level synthesis estimators are needed to approximate the hardware values $(ha_i$ and $ht_i)$. Additional data (the average number of times the task i is executed, n_i, and the average number of times communication between nodes i and j takes place, $n_{i,j})$. are provided by a system behavior simulator. Communication related data $(comm_{ij}$, $t_{trans(i,j)}$ and $t_{synch(i,j)})$ are obtained by hardware-software co-estimation. Figure 3.2. shows the information flow around SHAPES. More details about this point can be found in [CLL96].

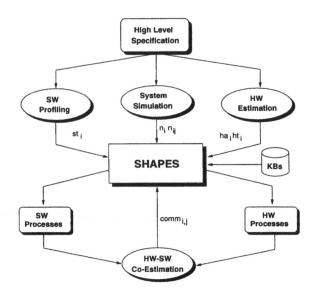

Figure 3.2. Flow of information within SHAPES

3.5.3 A Fuzzy Logic Expert System Architecture

The architecture of SHAPES is described in Figure 3.3. This block diagram describes the design flow within the expert system. The partitioning process starts with an initial *classification* phase. The result of this stage is a new attribute in the nodes of the task graph: the implementation degree. Based on this attribute, the *assignment* of tasks to hardware or software is performed. This assignment is constraint-oriented: blocks are assigned to hardware or software depending on the critical constraint of this particular design (basically, hardware area or system performance). The quality and suitability of the proposed partition is checked out in the *evaluation* phase. If the evaluation of the partition is not satisfactory, a new partition is considered following a particular *strategy*. A special procedure within the strategy module is the communication based *reordering*, which modifies the classification results according to the communication penalty in the hardware-software boundary. This strategy is selected from a knowledge base of tentatives. These modules and their implementation will be explained in the following sections.

3.5.3.1 Classification Module

The first step in a system partitioning process is to determine which blocks of the initial specification are more qualified to be implemented in special purpose hardware and which blocks are more suited to be implemented as software running on a standard processor. That step is performed in SHAPES by a rule based classifier module. Fuzzy variables are used to specify the rules.

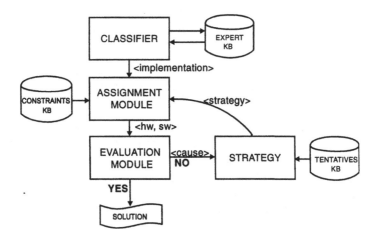

Figure 3.3. Block Diagram of the Expert System

The output of the classifier is the *implementation degree* of every task, a parameter that indicates the tendency of the task to be implemented either as hardware (close to zero) or as software (close to one). Figure 3.4 shows the linguistic variable associated to the *implementation degree* parameter. A detailed explanation of the meaning of this plot can be found in Appendix A.

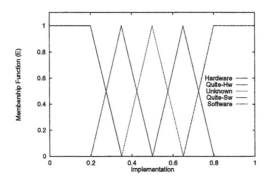

Figure 3.4. The linguistic variable *Implementation = [Hardware, Quite-Hardware, Unknown, Quite-software, Software]*

The classification module (as well as the rest of knowledge based modules that constitute SHAPES) is composed of three stages:

1. A *Fuzzification* stage that maps "crisp" input data into fuzzy sets. In the classifier module three fuzzy sets have been defined: `hardware_area`, `time_improvement`, and `number_of_executions` (the number of times a task is executed), each one corresponding to the information stored in the task

graph. Input data are thus mapped into linguistic variables on these sets, and can take values such as "large area", "high time improvement" or "many execution times".

2. An *Inference* process that applies the rules defined in the knowledge base on the fuzzified input data to make inferences and draw conclusions. In this work, the Fuzzy_CLIPS [FZ95] inference engine has been used.

3. A *Defuzzification* stage that maps fuzzy sets (the outcome of the fuzzy inference process) into crisp numbers. The defuzzification method that has been used is the *Center of Gravity Algorithm* [Me95]. It is needed to generate the output results that will be used in "non-fuzzy" tools (scheduler, estimation, etc.), although fuzzy values are kept in the memory of the system to be used in later steps. In the case described here, a crisp value of implementation is produced to obtain an initial partition.

The output of the classifier block is the set of input tasks ordered by their implementation degree. Figure 3.5 shows the ordered set produced when a specification example with 23 tasks is entered into the module. If no rule is activated for a given input task, the implementation degree is unknown (0.5). Furthermore, in that case, the tool can be configured to ask the designer (a supposed expert) the implementation value of the specific task. That information can be included in the knowledge base.

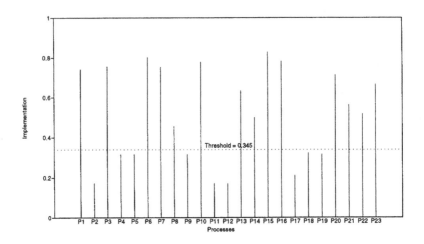

Figure 3.5. Classification results for the proposed example

3.5.3.2 Assignment Module

Once all the nodes have been classified according to their implementation degree the assignment module allocates part of them to hardware and the rest to software.

This module takes as inputs the ordered set of tasks and the system constraints (maximum allowed hardware area, A, and maximum allowed execution time, T), and produces as output another parameter, the *threshold*, which determines the hardware-software boundary (see Figure 3.5).

The knowledge base of this module holds a set of rules that characterizes the specification constraints to determine a reliable partition. The output data (a crisp threshold value) is obtained after estimating the "hardness" of the specification requirements: how critical (or not) the constraints are regarding the extreme values of the system performance. Two fuzzy variables are then defined: `time_constraint` and `area_constraint`. Both are dynamically defined because their values depend on the particular input data (the universe of discourse of the linguistic variable changes with the system under evaluation). For instance, `time_constraint`, which can be seen in Figure 3.6, is defined as a function of the *Time Constraint Ratio*, ρ_T,

$$\rho_r = \frac{T}{MaxT} \quad with \quad \frac{MinT}{MaxT} \leq \rho_r \leq 1$$

where T is the allowed execution time for the particular system being partitioned, $MinT$ is the execution time for the all hardware solution, and $MaxT$ is the execution time for the all software solution. All these values differ for every specific case, this being the reason why the fuzzy variables must be dynamically redefined for every system to be partitioned.

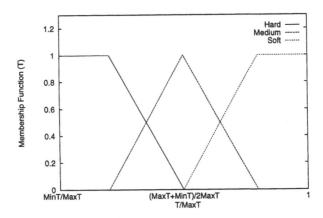

Figure 3.6. Membership Function of *Time-constraint = [Hard, Medium, Soft]*

3.5.3.3 Partition Evaluation

The value of the partition obtained by the assignment stage is measured by the evaluation module. In that module three parameters have been considered:

1. The estimation of the area needed to implement the hardware part A_p.

2. The scheduling of the assigned task graph, which gives the final execution time T_p.

3. The computation of the final communication cost.

For the hardware modules a rough area estimation is performed. Total area is computed adding the area estimates of all the blocks. Area estimates correspond only to one of the possible points of the design space for a hardware block. So far, resource sharing is not considered, but work is currently focused on including it in the area estimation, and choosing the most suitable point of the design space for each hardware block.

Task scheduling is implemented by means of a list-based scheduling algorithm. The scheduler takes into account the timing estimates of every task in the graph and the dependencies between them. As output it gives the final execution time for that partition, T_p, and the communication cost produced in the hardware-software interface. Again, timing estimates correspond to an average point of the design space. Tasks can be executed concurrently both in the standard processor and the application specific co-processor (ASIC).

The communication cost is composed of two factors: the total number of transfers that occur in the hardware-software interface, and the *communication penalty*, that is the global delay introduced in the system latency due to specific communication waiting.

Once those parameters have been calculated, the evaluation module determines if the proposed partition is feasible. If not, the strategy module must start working to refine the partition. Extra data with information about the "state" of the constraints (if they have been met, the existing gap, if not, the degree of violation) is then originated. Specifically, the following variables are defined:

*   The area and time overhead (with its linguistic variables `area_gap` and `time_gap`): $\Delta A = A - A_p$, and $\Delta T = T - T_p$, being A_p and T_p the current partition area and latency.

*   The communication penalty, that will be used to compute the *communication rate*, (section 3.5.3.5).

*   The processor and ASIC throughputs (fuzzy variables `proc_thr` and `asic_thr`) τ_{proc} and τ_{asic}, defined as the relationship between the execution time of the software or the hardware parts and the global system latency, respectively:

$$\tau_{proc} = \frac{\sum_{i \in SW} st_i}{T_p} \qquad \tau_{asic} = \frac{\sum_{i \in HW} ht_i}{T_p}$$

3.5.3.4 Strategy Module

When a partition is not feasible, another partition must be built upon the initial one. The threshold needs to be tuned, and it may be necessary to swap the tasks around the hardware-software boundary. To perform all these operations a particular strategy must be chosen. This strategy has to take into account the reasons why the partition is not correct. The strategy selection has been implemented with another knowledge based module. In this case, the rule base contains the usual tentatives that an experienced designer follows to improve the final system performance. The module takes into account the data generated by the evaluation module and, founded on the knowledge base, two non-exclusive alternative strategies can be selected:

1. To migrate a task from hardware to software or vice versa according to the constraints. This is carried out by tuning the *threshold* obtained in the assignment phase.

2. To improve the partition trying to minimize the communication on the hardware-software boundary (it relies on the presence of communication penalty and a balanced partition, i.e., $\tau_{proc} / \tau_{asic} \cong 1$)

The output of that module is fedback to the assignment module, and the partitioning process starts again. This iteration is performed until either a feasible partition is found or the strategy module determines that no solution can be obtained with the initial constraints. In this case, SHAPES asks the designer to change the system requirements and gives advice on how to do it. Otherwise, no solution can be given.

Communication based Reordering

After obtaining the first partition evaluation, SHAPES estimates the communication overhead by looking for the tasks that introduce a communication penalty, and by modifying their implementation value. In fact, a shift operation is performed in the ranking of tasks according to the *Communication Rate* factor, ρ_{comm}. This can be computed for both hardware and software tasks and relates the communications of a hardware (software) task i with all the software (hardware) tasks, with the total communications of the task i. This gives a measure of how *attracted* a task i is to the hardware or software sides regarding communications. The *Communication Rate* factor of a software task is calculated as

$$\rho_{comm}^{HW}(i) = \frac{\sum_{j \in HW} t_{comm}(i,j)}{\sum_{j \in HW} t_{comm}(i,j) + \sum_{k \in SW} t_{comm}(i,k)} \quad with \quad 0 \le \rho_{comm}^{HW}(i) \le 1$$

It considers the communications of the task i (implemented in software) with the tasks placed in hardware, being relative to the its total number of communications. In the same way, ρ_{comm}^{SW} can be calculated for tasks implemented in hardware. Note that $t_{comm}(i,j)$ equals zero if there are no data dependencies between tasks i and j.

The factor ρ_{comm} is estimated and fuzzified for all the tasks that lead to a delay (a communication penalty) in the hardware-software interface. A new linguistic variable is defined, the comm_shift, that has the attributes {shift_hw, balanced or shift_sw}. Taking into account the new data and the fuzzy implementation values computed in the first module and stored in the memory of SHAPES, the tasks affected by communication penalty update their implementation value. If the communication factor is high and the task is close to the threshold, the kind of implementation might even be changed.

3.5.4 Tracing an Example

This section develops an example to illustrate the way SHAPES works. The selected example consists on a task flow graph with 23 nodes whose initial classification has already been shown in Figure 3.5. This classification is based on estimates previously attached to the nodes and edges of the task graph. For that graph, system requirements are bounded by the values:

- All software execution time $MaxT = 1276$ t.u. (time units),

- All hardware execution time $MinT = 460$ t.u., and

- All hardware solution area $MaxA = 41,570$ e.g. (equivalent gates).

Constraints		First iteration results			
A	T	Threshold	Estim. Area	Estim. Time	Iterations
40000	1200	0.15	0	1276	2
40000	500	0.69	28270	504	3
30000	1000	0.28	7640	1017	2
30000	500	0.69	28270	504	3
20000	1200	0.21	6110	11088	1
20000	800	0.43	20450	545	2
10000	1200	0.14	0	1276	2
10000	800	0.31	7640	1017	-

Table 3.2. Results obtained with the proposed case study for different system requirements.

When entering different constraint values (hardware area *A* and maximum execution time *T*) the results obtained in the first iteration can be seen in table 3.2. In this table the different input constraints that have been applied to the example are shown (two first columns). The threshold column shows the value returned by the assignment module. This value can give an idea of how "hard" or "soft" the constraints are, since it indicates the "amount of blocks that need to be implemented in hardware". Values close to 0 result in "very software" solutions while values close to 1 illustrate the contrary situation. The columns with estimates reveal the partial results provided by SHAPES after the first iteration of the partitioning. The last column shows the number of iterations needed to get a feasible partition. A hyphen in that column means that there is no solution for those constraints. A comparison with other partitioning methods would be of interest in order to set up the worth of the partitioning strategy discussed here. However, the absence of co-design benchmarks makes this task quite complex. In general, different co-design environments target different system architectures, types of applications, ... and assume a variety of specification models oriented to obtain particular data closely tight to the chosen partitioning strategy (i.e., coarse or fine granularity). This makes it unfeasible to obtain examples that produce reliable comparison data.

In our case, analyzing the obtained results for the previous example (and with others not reported here), it can be said that SHAPES produces good results in terms of execution time. If the specification constraints can be met, the expert system can find a reasonable solution after very few iterations, what results in very short computation time (just a few seconds for big input graphs).

3.6 Conclusions

This chapter has presented one of the main tasks when developing complex and heterogeneous systems: the hardware-software partitioning phase. First of all, the importance of partitioning within hardware-software co-design has been justified. After that, several important issues have been introduced to characterize the problem, like the system model defined for partitioning, the different costs that can be considered to guide the process, the target architecture into which the system will be mapped or the application domain considered. These issues have been identified and briefly analyzed for the most well-known approaches and a method based on artificial intelligence techniques has been then presented.

The partitioning method introduced in this chapter presents several advantages. First, the use of fuzzy magnitudes allows us to handle imprecise and uncertain information (high level system characterization and estimation) in a clear and easy way. Second, the utilization of an expert system permits us to take advantage of the knowledge accumulated by system designers. Finally, it produces interesting information to the user, supporting designer interaction.

3.A Appendix

A *crisp set* A in a universe of discourse U can be defined by listing all its members $A = \{x_1, x_2, ...x_n\}$, where $x_i \in U$. This can be specified by a zero-one *membership function* for A, $\mu_A(x)$, such that

$$\mu_A(x) = \begin{cases} 1 & \text{if } x \in A \\ 0 & \text{if } x \notin A \end{cases}$$

A *Fuzzy Set F* defined on a universe of discourse U is characterized by a membership function $\mu_F(x)$ which has values in the interval [0,1]. $\mu_F(x)$ provides a measure of the similarity of an element in U to the fuzzy subset. It can be represented as:

$$F = \{(x, \mu_F(x)) / x \in U\}.$$

Set operations such as union, intersection or complementation have also been introduced to work with fuzzy sets. For instance, very often these operations can be defined in the following way:

$$\mu_{A \cap B}(x) = min \ \{\mu_A(x), \ \mu_B(x)\}$$

$$\mu_{A \cup B}(x) = max \ \{\mu_A(x), \ \mu_B(x)\}$$

$$\mu_{\overline{A}}(x) = 1 - \mu_A(x)$$

A *linguistic variable* is characterized by a quintuple $(x, T(x), U, G, M)$, where x is the name of the variable; $T(x)$ is the term set of x, i.e., the set of names of linguistic values of x, with each value being a fuzzy variable denoted generically by x and ranging over U. U is the universe of discourse; G is a syntactic rule (usually with the form of a grammar) for generating the name X of values of x. M is a semantic rule for associating with each X its meaning, $M(X)$, which is a fuzzy subset of U.

For example, let's have a look to linguistic variable shown in Figure 3.4. For this variable x = *Implementation degree*; the set of linguistic values of *Implementation degree*, according to the syntactic rule G, may be defined as $T(x)$ = *{Hardware, Quite-Hardware, Unknown, Quite-software, and Software}*. The universe of discourse U is the possible range of the implementation value (in this case [0,1]). $M(X)$ defines a fuzzy set for each term $X \in T(x)$.

Linguistic variables are the base for approximate reasoning. The process of extracting conclusions from a set of fuzzy antecedents is called *fuzzy inference*. Fuzzy inference is based on a set of decision rules whose antecedents and consequents are linguistic terms:

```
IF the task area is NOT large AND has good time improvement
THEN the task implementation is Quite-Hardware
```

These are very basic concepts. For more information about the use of fuzzy logic in expert systems and engineering applications the reader is referred to the classical literature, for instance [Me95, Zim91].

Acknowledgments

The authors would like to thank C.A. Iglesias for his insights in artificial intelligence techniques and for his fruitful collaboration.

References

[BML97] I. Bolsens, H. de Man, B. Lin, K. van Rompaey, S. Vercauteren and D. Verkest, *"Hardware/Software Co-design of Digital Communication Systems"*, Proceedings of the IEEE, vol. 85, 3, pp. 391-418. 1997.

[CGJ94] M. Chiodo, P. Giusto, A. Jurecska, H. C. Hsieh, A. Sangiovanni-Vincentelli and L. Lavagno, *"Hardware-Software Codesign of Embedded Systems"*, IEEE Micro, pp. 26 36, Aug. 1994.

[CLL96] C. Carreras, J.C. López, M.L. López, C. Delgado, N. Martínez, L. Sánchez. "A Co-Design Methodology Based on Formal Specification and High-level Estimation". *Proc. 4th Int. Workshop on HW-SW Codesign*, pp. 28-35. March 1996.

[COB95] P. Chou, R. B. Ortega and G. Borriello, "The Chinook Hardware/Software Co-Synthesis System", *Proc. 8th Int. Symposium on System Synthesis*, Sep. 1995.

[EHB93] R. Ernst, J. Henkel and T. Benner, *"Hardware-Software Cosynthesis for Microcontrollers"*, IEEE Design & Test of Computers, pp. 64-75, Dec. 1993.

[EPK97] P. Eles, Z. Peng, K. Kuchcinski and A. Doboli, *"System Level Hardware/Software Partitioning based on Simulated Annealing and Tabu Search"*, Journal of Design Automation for Embedded Systems, vol 2, No 1, pp. 5-32, 1997.

[FK94] M. Fares, B. Kaminsa. *"Exploring The Test Space With Fuzzy Decision Making"*. IEEE Design & Test of Computers, pp. 17-27, Fall 1994.

[FZ95] R. A. Orchard. *"FuzzyCLIPS Version 6.04. User's Guide"*. June 1995.

[GTW93] E. Glover, D. Taillard and D. de Werra, *"A User's Guide to Tabu Search"*. Annals of Operations Research, vol 41, pp. 3-28, 1993.

[GM93] R.K. Gupta, and G. de Micheli. *"Hardware-Software Cosynthesis for Digital Systems"*. IEEE Design and Test of Computers, pp. 29-41, Sept. 1993.

[HE97] J. Henkel, R. Ernst. "A Hardware/Software Partitioner using dynamically determined Granularity". *Proc. DAC*, pp. 691-696, 1997.

[KGV83] S. Kirpatrick, C.D. Gelatt and M.P. Vecchi, *"Optimization by Simulated Annealing"*, Science, vol. 220, number 4598. 1983.

[KL70] B.W Kernighan, S. Lin. *"An efficient heuristic procedure for partitioning graphs"*. Bell Syst. Tech. J. vol. 4. n. 2, pp 291-308, 1970.

[KL94] A. Kalavade, E. A. Lee. "A Global Criticality/Local Phase driven Algorithm for the constrained Hardware/Software Partitioning Problem". *Proc. 3th Int. Workshop on Hardware/Software Codesign*. 1994.

[KL97] A. Kalavade, E. A. Lee. *"The Extended Partitioning Problem: Hardware/ Software Mapping Scheduling and Implementation-Bin Selection"*. Journal of Design Automation for Embedded Systems, vol. 2, No. 2, pp. 125-164. 1997.

[Lag89] E. D. Lagnese. *"Architectural partitioning for system level design"*. Ph. D. ECE Dept., Carnegie Mellon Univ. Apr. 1989.

[LCL96] M. L. López-Vallejo, C. Carreras, J. C. López and L. Sánchez. "Coarse Grain Partitioning for Hardware-Software Codesign". *Proceedings Euromicro 96*, Praga, Sept 1996.

[Me95] J.M. Mendel. "Fuzzy Logic Systems for Engineering: a Tutorial". Proceedings of the IEEE. Vol. 83, no. 3, pp 345-377. March 1995.

[MGK97] J. Madsen, J. Grode, P.V. Knudsen, M.E. Peterson and M.E. Haxthausen. *"LYCOS: the Lyngby Co-Synthesis System"*. Journal on Design Automation for Embedded Systems, vol. 2, No. 2, pp 195-236. 1997.

[TCF96] A. Trouble, J. Chávez, L. G. Franquelo. *"FASY: A Fuzzy Based Tool for Analog Synthesis"*. IEEE Trans. on CAD, no. 7, pp. 705-714, July 1996.

[VG95] F. Vahid and D. D. Gajski, "SLIF: A Specification-Level Intermediate Format for System Design", *Proc. EDAC*, 1995.

[VL96] F. Vahid and T. Le, "Towards a Model for Hardware and Software Functional Partitioning", *Proc. Workshop on HW/SW Co-Design*, pp. 116-123, 1996.

[VL97] F. Vahid and T. Le, *"Extending the Kernighan/Lin Heuristic for Hardware and Software Functional Partitioning"*, Journal of Design Automation for Embedded Systems, vol. 2, number 2, March pp. 237-261. 1997.

[Wo97] W. H. Wolf, *"An Architectural Co-Synthesis Algorithm for Distributed, Embedded Computing Systems"*, IEEE Trans. on VLSI Systems, vol. 5, No 2, pp. 218- 229, June 1997.

[Za73] L.A. Zadeh. *"Outline of a new approach to the analysis of complex systems and decision processes"*. IEEE Trans. on System Man and Cibernetics, pp. 28-44, Jan. 1973.

[Zim91] H. J. Zimmermann, *"Fuzzy Set Theory and its Applications"*. Kluwer Academic Publishers, 2nd. Edition. 1991.

4 AN INDUSTRIAL CASE STUDY IN HW-SW CO-DESIGN USING CASTLE

Paul G. Plöger
Horst Günther
Eduard Moser

4.1 Introduction

Co-design is concerned with the joint design of hardware and software making up an embedded computer system [Wol94]. A top down design flow for an embedded system begins with a system specification. If it is executable, it may be used for simulation, system verification or to identify algorithmical bottlenecks. In contrast to other chapters of this book, the specification is not developed in this case study, rather it is given from the beginning. Furthermore we are not concerned with partitioning or synthesis of dedicated HW. Instead we focus on the problem how to find an off-the-shelf micro-controller which implements the desired functionality and meets all specification constraints. If feasible, this is usually much cheaper then using dedicated hardware. This chapter will answer the question of feasibility for a real life problem from automobile industry.

Based on the application example of an electronic diesel injection controller (EDC), we present various tools of our co-design workbench CASTLE. Robert Bosch GmbH kindly provided the executable specification of an EDC. It was available partly as written documentation, VHDL code and a large C program. This executable simulates the controller behavior while interacting with a model of the environment of the system. Thus the program contained functions implementing the controller as well as functions simulating the environment and the communication of both parts. It originates from simulation studies carried out on an in-house parallel simulation system with 4 transputers [ETAS97]. This high level simulation ensures stability of the control behavior of the algorithm under varying technical parameters.

J.C. López et al. (eds.), Advanced Techniques for Embedded Systems Design and Test, 77-101.
© 1998 *Kluwer Academic Publishers.*

Specifying the controller as an executable C program releases the designer almost completely from considering specific details or restrictions of a target hardware architecture during an early design phase. To develop a cost-effective real-time implementation of the EDC in the light of HW/SW co-design, we will discuss the impacts of software variations like C program transformations and the impacts of hardware variations like using alternative micro-controllers. The initial program was neither optimized with respect to runtime, nor storage requirements, nor computing accuracy or ease of implementation on a micro-controller, so both aspects and their especially their interrelation needed closer attention. Compiling the available C program for a target processor does not lead directly to a technically or commercially useful solution. The compiler may exploit some processor specifics, but the effect is only limited. Accomplishing the required short computing time still remains a challenge.

Some CASTLE tools are applied to this system specification and they produced a detailed time and space analysis. We consider different implementations, algorithmic optimizations and different choices of hardware. The main steps in the design flow may be summarized as follows:

- Application of given input stimuli in order to detect performance hot spots and insure a good test coverage,

- runtime analysis of given algorithms,

- performance prediction on different hardware alternatives,

- optimization of algorithms with respect to runtime and/or space consumption,

- alteration of the specification by other rates of calculation,

- conversion of floating point computation into fixed point computation and

- storage requirement assessment.

Due to the lack of space the last two points are not covered here, see [The96]. Furthermore some transformations like encapsulation, modularization or streamlining communication between modules by removing global variables were applied prior to our tools and are explained only very briefly. The examination method of this case study is guided by well-known principles of software design. Different implementations of the given specification had to be compared to determine which alternative suits a specific hardware best. These alternatives have been developed manually since automated optimizers deliver only local improvements, but fail to produce different algorithms.

The rest or this chapter is organized as follows: section 2 briefly describes interfaces and main function of the diesel control , section 3 gives an overview of the CASTLE tools used in this case study. The main results and conclusions are presented in section 4 and we finish with a summary in section 5.

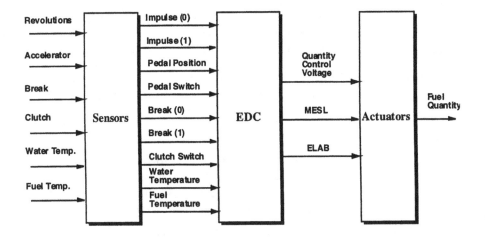

Figure 4.1. Block diagram of the fuel control process

4.2 An Electronic Diesel Controller as Application Example

This section presents an overview how the EDC is integrated into the whole fuel control process. In 1983 Robert Bosch GmbH together with Daimler Benz presented the concept of an electronic diesel control for truck engines for the first time at the international car fair in Frankfurt [Kor84]. The control has to compute the fuel quantity to be injected into the corresponding cylinder for every working stroke of the engine (combustion stroke of a cylinder). Computation depends on operating parameters which are measured by sensors.

The EDC mimics hydraulic-mechanical, i.e. analog control processes by means of programmed digital electronics. It extends their functional scope and enables an adaptation of the control algorithm to the operating conditions of the engine by modifying stored parameters without requiring a change of mechanical components. Program and parameters are stored in an executable read-only memory [Kir88]. The general structure of the control system is depicted in Figure 4.1. The main inputs for the fuel control process are:

- The number of revolutions per minute (rpm) of the motor. Two sensors are used which transform the number of revolutions into sequences of impulses (Impulse(0), Impulse(1)). This allows to cross-check the sensor signals against each other in order to determine failures of the sensors.

- The accelerator position which is translated into a voltage indicating the position (Pedal-Position, a potentiometer voltage value - scaled from 0 to 99), and a boolean value (Pedal-Switch) which shows if the accelerator pedal is pressed.

- The break pedal position which is transformed into two sensor signals (Break(0), Break(1)) indicating whether the break is pressed, or not. The following states are possible: not applied/applied/excessively applied.

- The clutch state (pressed, not-pressed).

- The water temperature is determined by means of NTC resistance transmitters with an accuracy of 1 degree Celsius.

- The fuel temperature.

The sensors transform their respective inputs into electrical input signals for the EDC. Based on this input data a Quantity Control Voltage is generated which controls the amount of fuel injected into the engine. In addition, two security flags allow to switch off the output amplifier or the output valve, respectively.

The EDC itself can be subdivided into three parts, not shown in Figure 4.1. First, a signal-transformation-and-check-unit maps sensor signals (voltages or impulses) into variable values which are used during the process control calculation. This unit includes an appropriate scaling of the voltages and a transformation of the impulse sequences from the revolution sensors into the corresponding number of revolutions. These values are also checked for plausibility, e.g. if they stay in the expected range or if the two impulse sequences of the revolution sensors compare OK. In this way failures in the sensors are detected and the system can react accordingly.

Second part is the central-safety-system which reads the defect status of the various sensors, and sets safety flags accordingly. Three flags are used internally to reduce the output quantity, and two external flags (MESL, ELAB) directly influence the actuator.

Finally the fuel-quantity-control-unit inside EDC performs the main calculation algorithm. It can generate different quantity values which correspond to different operating conditions of the motor, namely for the start phase, for idle engine, for maximum load of the engine, for maximum number of revolutions, and for normal driving conditions. The suitable quantity for the current driving conditions is selected and transformed into the Quantity Control Voltage.

The information delivered by the sensors is compared with values permissible for the operating state. A request for increased performance is only fulfilled if this is feasible without exceeding the permissible operating values, e.g. the permissible maximum torque for a given engine speed. If an operating malfunction is recognized, the whole engine falls back to fail-save state like a complete engine turn off. This fault is displayed at the dashboard and stored in the controller for diagnosis during maintenance.

Bosch provided the control behavior in the form of a C program which contained only a minor part of the control as described above namely fuel-quantity-control-unit [Rei92]. Other time-critical processes, such as injection time control are not considered. Many continuous quantities are mapped onto floating point numbers in the C program. It was modularized according to the block diagram in Figure 4.1 and consisted of the following modules:

- A timer module containing global timers. Conditions like timing intervals for signal arrival or occurrence of specific events are defined here. If they fail, an error is assumed and the program branches to an error state. Essentially timing conditions synchronize the computing process of the controller with the technical process of the working engine.

- The interface module converts input values into an internal representation and scales them. Output values are transformed from the internal representation into external signals.

- The quantity control module computes the fuel amount considering the status of environment and sensors. In the case of recognized (or supposed) errors, uncritical fallback values are used or fuel supply is disconnected. The (simulated) sensors are sampled with a fixed frequency of 1 kHz and produce values for brake state, air temperature and cooling water temperature, accelerator pedal position and pressures. Rpm detection is different, see below, and clutch position is recorded only every 50 ms. Each sensor value is bound to a C variable.

For rpm calculation signals are triggered by a Hall sensor when the magnetized teeth of a revolving wheel mounted on the camshaft pass it. The revolving wheel delivers (2* number of the teeth / number of the cylinders) signals per working stroke. From the specification we concluded that a hardware device samples this signal with 100 kHz and stores the value in a register which is read periodically every 1 msec. From the delta time of two successive reads and the number of teeth, the rpm is calculated. Errors are avoided by using two revolving wheels and two signaling transmitters delivering independent rpm values. The two rpm values are checked for plausibility which adds fault-tolerance to the EDC.

The C program together with the written documentation unraveled specification constraints far beyond the intrinsic fuel calculation algorithm of the EDC. These decisively determine the final implementation and the selection of a micro-controller. By their specific nature, these constraints are difficult to handle automatically. Instead the designer must provide design alternatives of varying functionality and quality. It falls into her responsibility to define, e.g., which floating point variables should be converted into fixed-point variables, how to handle synchronization or which alternative algorithm implementations are available. Tools can assist during implementation of variants, e.g., conversion to fixed-point representation can be automated in principle. The impacts of such a conversion can again be check semi-automatically by simulation runs indicating whether the new variables exceed their permissible fixed point range and how much they accelerate runtimes compared to floating point computation.

Given the C program, it was difficult for us to identify the constraints for synchronization of technical process and electronic controller expressed implicitly in a C program. Synchronization constraints and input/output rates of signals might better be expressed in other languages which address synchronization problems more explicitly.

4.3 The CASTLE Workbench

Co-design tools cover a considerable span of topics due to the wide spectrum of embedded applications. E.g. selection of an off-the-shelf standard micro-controller from a number of different processor families is at one end, designing a dedicated chip for a time-critical task at the other extreme. Taking this fact into account we decided that it is impossible to answer all occurring questions with one single monolithic tool. Therefore, CASTLE was planned as an extensible tool suite based on a common internal format SIR (System Intermediate Representation). High-level programming languages like C/C++, and hardware description languages like Verilog and VHDL 93 are parsed to this common format. It is based on a C++ class library of persistent objects. Control and data flow are mapped to graphs, data types to an extensible tree. Data structures can be visualized and processed in this format, typically by annotation with task-specific data. E.g. the profiler Kukl (see below) annotates a system description in SIR with counters which trace execution frequencies. Output of CASTLE can either be in high-level programming languages like C or VHDL, or on lower description levels like sequential circuits or net-lists in BLIF and structural VHDL. This is achieved after a simple minded internal synthesis. The C++ class library and the tools of CASTLE are freely available in the source code for some UNIX systems (HP, DEC, SUN, Linux).

The co-design approach pursued in CASTLE is designer-centered, i.e. it is the designer who definitely decides the design. The tools do not make independent decisions on important problems but only provide measured values for supporting decision making. In the following we outline briefly the functionality of CASTLE tools used for EDC co-design. We begin with an overall picture of the performance, carry on with a coarse grained performance prediction for different micro-controllers and finish with a detailed analysis of the final application code compiled for the target processor. Dynamic and static data size consumption and elimination of floating point variables are final steps carried out on SIR.

4.3.1 Procov

Procov is an abbreviation for 'program coverage'. It closely resembles the tcof program found on some UNIX systems. In contrast to it, it is portable since it only requires a GNU compiler, which is ubiquitously available. The application program to be analyzed is first compiled with a special option, then runs on some sample input. Every count obtained during profiling refers actually to a piece of straight line code, called a basic block. Usually new basic blocks are generated by the compiler at if- or then-statements, at loop bodies or short-circuit operators ('&&' or '||'). These operators show that it is possible to have more then one basic block produced for a single source line of code. In the case, the maximum execution frequency of all basic blocks associated to this line is displayed. Data is collected in a profile, Procov reads it and produces as global output a list of most frequently called functions and as local output the most important program lines of each source file. Global output lines are sorted according to execution count and line numbers. Furthermore the corresponding function names are given, Table 4.3 shows a summary view of Procov running on Sparc. The local analysis in Table 4.1 reveals, which lines have been executed at all and which of them are most frequently used. Most likely the line containing the peak count points to a performance hot spot, though this number indicates just a count not a runtime. We see that the while loop is executed on average 499791/150003 = 3.3 times per call. However, the display is coarse grained because most frequently a program line is compiled into several instructions, which in turn can be distributed over different basic blocks due to compiler optimizations. E.g. Table 4.1 does not show how often the 'then' part and/or 'else' part of the 'if' condition was executed. Taking a closer look at assembler instructions explains that the else part was moved into the delay slot of the 'if' instruction. So it is not regarded as an independent basic block and consequently could not be annotated with a count in Procov output.

Count	Program line
static	`short *Finde_Intervall(`
	`short *iu,`
	`short *io,`
	`short x)`
	`{`
150003	`short *l = iu;`
	`short *r = io;`
	`short *m = l + (r-l)/2;`
150003	`while(l<m)`
	`{`
499791	`if (*m <= x)`
	`{ l = m; }`
	`else { r = m; }`
499791	`m = l + (r-l)/2`
	`}`
	`return m;`
	`}`

Table 4.1. Procov output for a C function

Two other annoyances may distort Procov outputs. Firstly, in global view, when the compiler optimizes a program and short functions get inlined, a source line may become associated with more then one function name. The inlined function was dissolved into several places, but the line information of the containing function is kept. As a consequence more the one function name may be associated to a single line in a file. Secondly, if preprocessor macros are used, the operations hidden behind the macro replacement cannot be mapped to C source text anymore.

The counts displayed by Procov in front of a program line specifies the maximum of the execution frequencies of all basic blocks which can be assigned to this line. The runtime of a single line is determined by the machine instructions assembled for this line. The total execution time of a line is the product of the execution time of machine instructions times execution frequency. In the case of a high execution frequency, this single line time should therefore also be checked. Because gcc may reorder statements, optimize them away or inline functions, and because functions may be split or hidden inside preprocessor macros, the given line execution frequency may not apply to all operations in that line. The best match between line frequency and program source is obtained when compiling without optimization and without using macros. Procov is ideal in getting a global overview of counts fast. Refined tools must applied to include basic block structure into the picture.

4.3.2 Kukl

Given a system implementation in C Kukl estimates runtimes on different candidate micro-controllers. The estimate is fast and yields results for all candidates in one sweep. To this end Kukl analyzes program parts without branches statically. To include the dynamic program behavior, Kukl inserts counting instructions into the control data flow. Thus it indicates in an early design phase which micro-controller meets the time requirements of the specification and/or which clock frequency a controller requires. As machine characteristics Kukl considers cycle time, instruction times, data width of operations and floating point coprocessor support. Peripheral devices of the controllers, prices, pipeline and cache effects remain unconsidered. Hard real-time requirements can be estimated plausibly but can not be validated.

To produce an estimation Kukl needs three different inputs: the program, the profile, and tables of instruction timings for the different micro-controllers. The workflow starts with compiling the implementation into the intermediate SIR format, see Figure 4.2. This format is a graph based representation of control and data flow of the source program. It is used as a generic assembler and is therefore independent of the specific features of a micro-controller. A one-to-one reconstruction of the program is possible. Kukl traverses SIR and inserts counters at each branch of the control flow, i.e. at all function calls, at all conditions and loops. Thus each essential basic block is traced. This annotated representation is transformed back into a C program, and gets compiled and linked. The new executable is run on typical input data. Since the annotation is done on the level of the generic system description and not with the instruction set of a specific controllers, it always delivers unique counts regardless of the type of machine used for profiling.

Code coverage is checked with Procov, as described in the previous section. Results of this profiling serve as the second input for Kukl. These profiles extend those used by other profilers like, e.g. gprof, which measure just the execution count of functions. SIR profiles additionally account for intra-functional branches like loop counts or the number of true evaluations for all conditions. In this respect our approach is a refinement of gprof. Another advantage lies in the fact that it definitely traces all function calls. This is sometimes a problem using gprof when functions are either to short to be detected or are inlined.

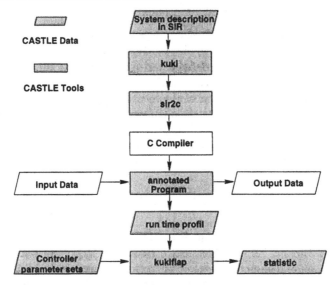

Figure 4.2. Steps of execution time estimation using Kukl

The remaining third input are tables of runtime intervals of generic SIR assembler operations on candidate controllers. The table entries contain an operation name, data width and minimal, nominal and maximal timing. In principle this timing information can be taken from data books of the respective manufacturer. This procedure is time consuming and error prone though. Besides that it falls short for those operations, which cannot be mapped to single instructions. Examples are basic arithmetic operations on a candidate micro-controller which has a smaller bus-with then data type of the operation. These operations must be implemented by library functions. Their timing is neither given in the hardware manual nor in the library. Another example is a conditional assignment or evaluation of logical expression trees from leaf expressions, where a sequence of instructions has to represent a single C operator. Direct measurement of dedicated small stopwatch programs are the best way to obtain these timings [Plo97].

Now the runtime estimation is a sums of products. For each basic block found in the SIR file its count it multiplied by the time of all operations in this basic block. These are specific for the candidate and may be intervals. Summation is taken over the execution path. We get a spectrum of minimal, typical and maximal times used for each basic block on each candidate controller. Worst case execution time can also be modeled. For this purpose, we assume for every branch of the control flow that the most time-consuming path is selected. For loops we sum up worst case loop repetition counts, which are supplied in the profiles. The estimate for a mean, nominal and longest program execution is determined by the same summation as before.

Observe, that these times account for a different uncertainty in the runtime estimate than the minimal, nominal and maximal times. The mean, nominal and longest numbers refer to different control flows possible, while the above mentioned intervals cover data depended runtimes of instructions.

Though being approximate Kukl estimates a great number of controllers fast. Since Kukl should be used early in the system design process where system specification is still coarse and not yet definite, the inaccuracy does not harm. For a more precise system description, the analysis accuracy must be refined by using tools such as Paladin though at the expense of a longer response time. Currently, Kukl has parameter sets for processors in Intel's x86 family as well as some Sparc processors. Further sets can be produced either manually on the basis of the instruction times from data books or semi-automatically by a number of instruction time measuring programs.

4.3.3 Paladin

After choosing a suitable candidate with Kukl, more detailed information can be obtained by Paladin. This profiler lists a statistic of static and dynamic behavior during program execution for a *specific* processor. So in contrast to Kukl we trade in genericity for accuracy. The program statistic contains the following information:

- instruction mix of compiled and executed code,

- usage of addressing modes,

- full blown address trace with read/write count of all registers and memory locations,

- data transfer analysis between arithmetic/logical units and registers,

- bit widths statistics for occurring constants,

- branch statistics,

- subroutines invocation and execution time calculation.

All these results are exact data for the specific processor. Therefore the prerequisites for using Paladin are much harder then in case of Kukl. We need a compiler which instruments the system description. This enables the measurement of the instruction frequencies when executing the instrumented program. Next, we need the target processor since the program only works for this particular processor type. CASTLE includes a modified GNU C++ compiler, called Gax, which instruments the produced code in the desired way. Gax and Paladin combinations are currently usable on Sparc, i80386, i80486 and mc680x0. An extension to other processors which are supported by the GNU compiler is possible without great difficulties.

The Paladin workflow, Figure 4.3, starts from a C/C++ implementation. Compilation with Gax instruments every basic block. Only the last instruction of a basic block can be a jump instruction. The instrumentation counts the execution frequency of each basic block. In addition, branch frequency and target are recorded, i.e. jumps executed from one basic block to another block. Similar to Kukl a profile is generated when executing the program which was instrumented with Gax. This runtime profile and the disassembled implementation to be analyzed are the inputs to Paladin. It then generates a detailed analysis of the program run.

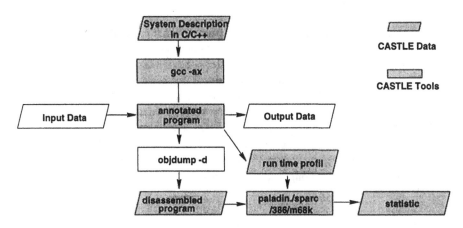

Figure 4.3. Analysis of system behavior with Paladin

E.g. logging the execution trace of all basic blocks is possible, which gives the exact sequence of program branches and counts of all loop bodies. However, this trace is fairly large since a new basic block begins on average after each 5 instructions and a processor can execute several millions of instructions per second. The amount of data should then be compressed by piping the runtime profile through GNU gzip. Usually, compression factors of 100 to 1000 can be achieved since sequence of visited basic blocks is very repetitive. The last step of disassembly is done by the GNU program objdump. The main advantage of Paladin is its ability to run the real code, i.e. including all conceivable compiler optimizations. Furthermore it is accurate down to the level of basic blocks which is as exact as one can get without a dedicated HW or emulation model of the target controller.

To sum up the functionality of the three CASTLE tools presented in this section, we note that the system provides analysis on different levels of abstraction. This can be used to refine the design space as follows: Procov can be used to identify hot spots and code coverage via execution counters. Then Kukl checks runtime requirements on a set of candidate controllers. After selecting a particular processor, Paladin analyzes and validates the performance on this candidate. All examinations are based on input data selected by the designer as being typical for the application. So far CASTLE cannot analyze code which has not been covered by input data.

Since the functional description of the EDC has been explained in the preceding section and tools in this one, all prerequisites are met to apply the tools to the specification and analyze the C programs of the controller.

4.4 EDC Analysis using CASTLE

This chapter shows how CASTLE tools are applied in the EDC design flow. The main problem studied in this section the selection of a suitable micro-controller and the necessity to enhance algorithm implementation to meet the given specification constraints.

After adapting the original EDC simulation, overall program performance is explored systematically by the CASTLE tools: Procov and Kukl. We take a closer look to the function encapsulating the controller behavior and compare it to different micro-controllers. Then we vary the algorithm of the most demanding function. A detailed analysis using Paladin gives insight to some co-design tradeoffs followed by a brief cache analysis. Finally we conclude by an investigation what can be gained for an EDC if we allow the variation of the initial specification but do not compromise in system behavior. All results are displayed and evaluated with regard to an implementation in software running on a standard hardware. This metric is a basis for the decision process among experts of different disciplines working on the design of the EDC.

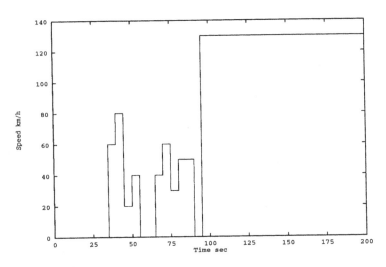

Figure 4.4. Standard driving curve

4.4.1 Preparation of the Experiments

The original version of the system description which GMD received from Bosch for this case study had two functions namely environment simulation and a second for the diesel injection control itself. It used floating point arithmetic for the most operations and is referred to as the floating point version. We separated the entangled parts manually into simulation and control part. The control part was re-written using only integer variables and operations. Furthermore, some functions were simplified. This version is referred to as the integer version. In the following the optimized versions of original functions are always prefixed with GMD. The kind of optimizations applied during the split are explained in more detail in the section optimization below. This final adapted description of the control part allowed direct compilation to an executable for a microcontroller just by changing some C preprocessor defines.

The main simulation loop is shown in Figure 4.5. In a car all tasks run in parallel, but for the C driven simulation they are sequentialized by a simple round robin schedule. The function Steuergeraet implements the EDC task and is triggered every one msec. All numerical results refer to a simulation period of Sim_Ende = 50 s if nothing else is mentioned. Standard simulation input is the speed curve shown in Figure 4.4.

```
while (Globtime <= Sim_Ende)
     Fahrer();
     Motor_Fahrzeug();
     Sensorik();
     Steuergeraet();
     Aktuatorik();
     timer_interrupt(dt);
}
```

Figure 4.5. Main simulation loop

Table 4.2 shows the abbreviations of the used set of candidate processors. These have been selected as to allow the data predicted by the CASTLE tools to be verified by executing the programs on available machines. No controller starter kits or development boards were available at the time of that experiment.

Name	Processor Type	Clock rate
386SX	i80386SX	25 MHz
486DX	i80486DX	33 MHz
IPC	LSI 64801Sparc	25 MHz

Table 4.2. Examined processors

When selecting a microcontroller we should also consider the compliance with the permissible service conditions, e.g. whether the permissible operating temperature range fits to the conditions in the vehicle. Furthermore an upper bound was defined for the sampling rate of the controller. For reasons of electromagnetic compatibility, the largest external frequency which occurs in the EDC must not exceed 25 MHz. So the 486DX is nearly out of range but yields important insight for the worst case execution time, see below.

4.4.2 Procov

We want to detect the performance hot spots. Table 4.3 summarizes results from two complete simulations of floating and fixed point versions respectively. Execution count, line number, file and function name are given. The right list containing functions of the fixed point version contains fewer items, since some of them, e.g. *Min, Max,* were eliminated during the preparation process mentioned in section 4.4.1. Others functions were optimized. For example, while *Motor_Start* is permanently active in the floating point version, it is used during start phase only in the fixed point version, so its execution count dropped significantly. We simulated using the standard driving curve of Figure 4.4. The total simulation time of 50 s corresponds to 50001 calls to *Steuergeraet*. Most likely code optimization pays off at source code lines with large relative frequencies. For example *Ipoll* in Table 4.4 has a code segment which is activated 1147141/50001 = 23 times more frequently than the rpm measurement in *Erfassung_Drehzahl.*

Floating Point version				Fixed point version			
Count	Line	in function	file	count	line	in function	file
1147141	203	*Ipoll*	edc.h	1360652	30	*GMDFind_Int.*	ipol.c
577948	300	*Ipol2*	edc.h	150003	111	*GMDIpol2*	ipol.c
499794	147	*Lin_Fkt*	edc.h	100002	56	*Erfassung_Drehz.*	Drehr.c
100060	72	*Max*	edc.h	50001	60	*GMDIpoll*	ipol.c
50001	94	*Min*	edc.h	50001	1139	*Menge*	menge.c
50001	1539	*Erfassung_Dreh.1*	steuer.h	50001	377	*Ueberw._Drehz.*	Drehr.c
50001	1822	*Erfassung_Dreh.2*	steuer.h	50001	210	*Erfassung_Ueber.*	steuer.c
50001	2280	*Erfassung_Ueberw.*	steuer.h	50001	237	*Steuergeraet*	steuer.c
50001	188	*Motor_Start*	steuer.h	50001	56	*Zentrale_Ueberw.*	steuer.c
50001	1489	*Steuergeraet*	steuer.h	...	...	...	...
50001	1444	*Zentrale_Ueberw.*	steuer.h	1036	219	*Motor_Start*	menge.c
...	...	...	...	...	...	...	...

Table 4.3. Line execution frequency of floating point and fixed point versions. Some functions were optimized away in the fixed point version

Table 4.4 shows a detailed Procov output for function *Ipoll.* Because it contains the highest absolute line execution count, it calls for special attention. This function searches for an interval of containment in a linear way followed by an interpolation. In the line with count 216 we see, that the rightmost extreme point is returned as interpolation value only 216 times during a total of 350007 calls. In all remaining executions of *Ipoll,* the inner search loop is passed on average $1.1 \ 10^6 /3.4 \ 10^5 = 3.28$ times per search call. Extending analysis to another function called *Ipol2* shows that it is called 150003 times and that it passes its inner search loop on average 577948/150003 = 3.85 times per call. Furthermore function *Ipol2* calls function *Ipoll* twice per call, in fact 300006 calls of *Ipoll* are initiated by *Ipol2.* Therefore, *Ipol2* is the most important candidate for improvements. We will return to this observation after we have checked the overall simulation time on our candidate processors using Kukl.

Count	Program	Comments
	`short Ipol1(IPOL1 *wp)`	
350007	`{`	
	`LIFU lifu_str; short *xwp, *ywp`	/* Pointer on X - Input /Y-Input values */
	`anz;`	/* Number of base points */
	`int i; xwp=wp->xp; ywp=wp->yp;`	
	`anz=wp->anz_x;`	
216	`if(wp->xw <= *xwp) return(*ywp);`	
349791	`if(wp->xw >= *(xwp+anz-1))`	
	`return(*(ywp+anz-1));`	
1147141	`for(i=0;i<anz;i++) {`	
1147141	`ywp++;`	
	`if(wp->xw <= *(++xwp)) {`	
349791	`if(wp->xw <= *xwp) {`	/* Search top & down limit value */
349791	`lifu_str.xu=*(xwp-1);`	/* ... fill linear structure */
	`lifu_str.xo=*xwp,`	
	`lifu_str.au=*(ywp-1);`	/* compute return value through Lin_Fkt */
	`lifu_str.ao=*ywp;`	
	`return(Lin_Fkt((LIFU *)&lifu_str,`	
	`wp->xw));`	
	`}}}`	/* for */
	`return 0;`	
350007	`}`	

Table 4.4. Procov output from the C function (Floating point version) with the highest line execution frequency

4.4.3 Kukl

Up to now we just checked execution counts but no runtimes. Kukl will identify suitable processors able to simulate the whole simulation in no longer then 50 s by runtime prediction. Meeting this timing constraint implies, such a processor can simulate in **real-time** which is a strong indicator that other timing constraints can be reached also. Following to this global consideration Kukl will validate the 1 msec constraint for the fuel quantity calculation in function *Steuergeraet*.

Along the x-axis of Figure 4.6 we have a group of columns called 'typical' containing a white bar in the middle, one for each of the 3 candidate processors from Table 4.2. Kukl predicted these runtimes for the candidates under the assumption of a typical control flow in contrast to a worst case control flow, see section 4.3.2. Neighboring columns of the white one denote estimations of runtimes for different data dependent instruction times. The white columns assume a nominal time for the execution of instructions and are called data nominal in Figure 4.6. The 386SX falls completely outside the real-time margin of 50 s, while IPC and 486DX easily master this limit. Other interesting conclusions may be drawn from this figure.

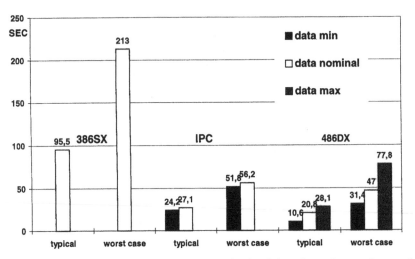

Figure 4.6. Runtimes of the C program on basis of data dependent and control flow dependent instruction times

4.4.3.1 Data-Dependent Runtimes

The execution times of C operations may vary according to processor and data they work on, like shift instructions on controllers without barrel shifter, multiplication instruction and operations on operands with different bit widths. Therefore, Kukl allows an interval for the execution time to be specified for each instruction. The bars labeled data max in Figure 4.6 show how execution time increases if data-dependent operations always take their maximal instruction time. We see a little variation of only 10% for IPC but up to 50% for 486DX. Sparc architectures are much more regular then the complicated Intel processors. As a consequence of this, instruction timings remain stable for all data types and almost all operations on IPC, while almost all of them vary with the data type and value on x86 processors.

4.4.3.2 Control-Flow-Dependent Runtimes

The length of the longest path is determined for a typical program run, e.g. by means of the standard driving curve. If the executed code contains loops with data depended bounds, the maximum of all encountered loop bounds is stored in the profiles. For worst case estimation we assume this loop bound to occur for every loop execution. In addition, we compute the most time consuming path for each program branch. Then Kukl calculates the runtimes on the assumption that this longest path is always taken. The control-flow-dependent runtimes are always added on top of data-dependent runtimes. These are the bars labeled worst case in Figure 4.6. We see that their variation has a more pronounced impact on total runtimes then the data dependent runtime intervals represented by pointed and grayed columns. For the 486DX we can make a remarkable observation. This candidate, which clearly masters the 50 s constraint under typical control flow and all considered instruction timings, will fail under the assumption of worst case control flow combined with worst case data timings. This proves, how easily all spare resources may get burned under worst case conditions. The diagram gives the designer a feeling of how much spare resources he should add to the design to be save even in the case of non-typical worst case conditions.

4.4.3.3 Machine-Dependent Runtimes

Cache and instruction pipelines in processors accelerate instruction execution. This fact can be addressed in Kukl by specifying a maximal data execution time for using no cache and a stalled pipeline and an execution time with cache and pipeline usage as nominal instruction time. Exactly this approach is used in Intel data books for the x86 family [Int92]. However, these effects can also be estimated in a separate step. In subsection 4.4.6 a simple way will be explained with function *GMDIpol2*.

4.4.3.4 Compiler-Dependent Runtimes

The execution time of a program is also dependent on the optimization quality of a compiler. For example, many compilers can transform a division by 2 into a shift instruction though hardly any compiler performs a simplification of common subexpressions beyond function boundaries. Such optimizations can also be carried out on SIR but have not been implemented yet.

We now focus on the function implementing the EDC behavior, namely *Steuergeraet* from Figure 4.5. This function is the essential component for quantity computation. It is called every millisecond and may therefore require no longer then one msec CPU time to complete. The compliance with this time limit is again examined with Kukl. Figure 4.7 presents the results for three different processors using the fixed point version of the function. The measured values are obtained from timing the simulation for the standard driving curve. Kukl determined runtimes for the tested processors which are up to 16% lower than the measured values. This accuracy is sufficient for a rough processor selection and estimation of the sampling rate. Kukl includes only a simple processor model. There are infinitely many registers according to this model, i.e. all variables are always located in registers. This neglects access times to variables in memory. Neither the length of the instruction pipeline nor the size of the cache are modeled. This also applies to operating system influences, interrupts or memory refreshs.

Our expectation from Figure 4.6 are confirmed. The smallest processor is not able to meet the time bound of one msec. The other processors are suitable for an EDC implementation. Kukl predictions are reasonably exact, although they underestimate the actual runtimes a bit. After identifying suitable candidates the next section accelerates code in *Steuergeraet*. Recall that this choice is based on the observation made at the end of section 4.4.2 by the help of Procov. It indicated that function *Ipol2* must be enhanced since it is called from *Steuergeraet* most often.

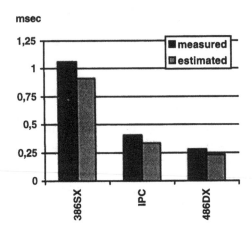

Figure 4.7. Estimated and measured run times for the normal case of the function *Steuergeraet*

4.4.4 Implementation Optimization

The analysis by means of Procov indicated that the two-dimensional interpolation is the most time-consuming part. Therefore, we now improve the interpolation calculation. Let a point (a,b) and a 2 dimensional table $T((X_0, Y_0), (X_1, Y_1),..., (X_M, Y_N))$ of non equidistant data points (X_n, Y_m) representing the function $F(x,y)$ be given. The interpolation function *Ipol2* calculates $F(a,b)$ in two steps: first it searches for two intervals i, j, which contain a and b, first two lines in function *GMDIpol2* in Figure 4.8. Second step are three one dimensional interpolations, i.e. function calls to *GMDIpol1* Our hand-optimized version is shown as indicated by the function prefix *GMD*. Our initial acceleration concerned the two function calls which search for an interval of containment for a and b. We replaced a linear search by a binary search using interval bisection and lowered the call count to *GMDFinde_Intervall* by some local code optimizations. This yielded about 18% less frequent execution of the while loop body in function *GMDFinde_Intervall*. Two more and differently implemented variants, called *BiLin* and *MyIpol2*, were investigated. Both use a slightly less optimal version of *GMDFinde_Intervall*, but inline the three function calls to *GMDIpol1* in the calling routine *GMDIpol2*. After some algebra 2 or 3 straight line expressions respectively remained. *MyIpol2* saves one division and several subtractions as compared to *BiLin*. They outperfomed the original *Ipol2* routine by 43% and 47%, but as the interval find function was less optimal as in *GMDFinde_interval* they could not reach the enhanced performance of the three GMD functions in Figure 4.8 which saved 57% runtime in total.

```
int GMDIpol2(int**T,int*X,int*Y,int a,int b){
    int i = GMDFinde_Intervall(X[0],X[N],a);
    int j = GMDFinde_Intervall(Y[0],Y[M],b);
    int tmp1 = GMDIpol1(a,X[i],X[i+1],
        T[X[i]][Y[j]],T[X[i+1]][Y[j]]);
    int tmp2 = GMDIpol1(a,X[i],X[i+1],
  T[X[i]][Y[j+1]],T[X[i+1]][Y[j+1]]);
    return GMDIpol1(b,Y[j],Y[j+1],tmp1,tmp2);
}
int GMDIpol1(int p,int low,int up,
    int lval,int uval){
        return low +
                (p-low)*(uval-lval)/(up-low);
}
int GMDFinde_intervall(int up,int low,int x){
    int l=low, r=up, m=l+(r-1)/2;
    while (l<m){
            if(m<=x){l=m}else{r=m};
            m=l+(r-1)/2;
    } return m;
}
```

Figure 4.8. Interpolation in 2 dimensions

This is how far we got just hand optimizing the code, not touching the data representation in T. To optimize the search time even further, we changed the data point table T. The characteristic diagrams have a technical-physical origin, so should be continuously differentiable. In this sense the linear interpolated values are false but obviously suitable for control though. Therefore, the question arises whether there are any other false, but equally suitable data point tables, which can be computed faster than the original one. We converted T to an equidistant representation T'. The implementation variant called *WIpol2* worked on T' and replaced two calls to *GMDFinde_intervall* by an algebraic expression evaluation which involves just one integer division. Addressing can be simplified even further, if the number of the points in the support of T' is a power 2. Now the integer division can replaced by a shift. We produced maximum error bounds of the output variable to check the correctness of the transformation of data and algorithms. We observed a maximal difference of 1.3%, the average deviation was 0.3%.

Optimization	Function	Kukl	timed	Kukl	timed	Kukl	timed
		IPC		486DX		386SX	
none	*Ipol2*	61	115	51,3	58,8	371	255,9
inlining	*BiLin*	53,9	65,5	35	36,4	178	116,9
inlining+no divide	*MyIpol2*	49,9	61	32,3	33,8	168	112,9
optm. Search	*GMDIpol2*	53,9	49	34,1	35,3	128	121,8
fitting data	*WIpol2*	34,6	37,2	20,2	24	66	75,3

Table 4.5. Time estimates (sec) of the two dimensional interpolation procedure

Table 4.5 summarize timed runtimes and Kukl predictions for the different variants. The timed vales are in seconds and refer to 1089000 calls to the interpolation function. The runtimes continuously improve with the new implementations on all computers. The fastest implementation *WIpol2* is by a minimum factor of 2.5 faster than the original implementation. Since this routine required approx. 25% of the total CPU time of the program, we obtain an overall acceleration of about 15%. However, the cost of this variant is a considerably increased data size for the enlarged equidistant data tables used by *WIpol2*, see section below, Figure 4.9. The estimated values in columns Kukl in Table 4.5 are by a maximum of 17% lower than the measured ones for 486DX. The other processors partly show even more significant variations, in positive and negative direction. Lower values are due to the already mentioned idealization of data traffic in Kukl. Higher values can be attributed to the fact that the measured values are based on the execution of a C program which was optimized during compilation. Kukl, however, determines the values by mapping the SIR description onto the specific processor without optimization.

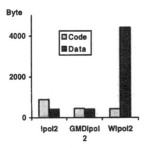

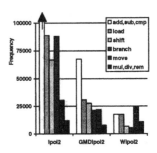

Figure 4.9. Memory requirements for three different variations of two dimensional interpolation

Figure 4.10. Execution rate for different instruction types using three variations of two dimensional interpolation

4.4.5 Paladin and Co-design Effects

We transformed and profiled the system specification and predicted performance on different hardware candidates. We now fix the IPC processor as target hardware and are still open for *GMDIpol2* or *WIpol2* as possible software variants. We keep *Ipol2* just as a reference. Programs are now compiled using Gax for Sparc which is also supported by a respective Paladin program. Figure 4.10 shows the instruction distribution for different instruction types on assembler level. This diagram includes all compiler optimizations, depicted is the spectrum of the final code actually running. From an instruction-set point of view *Ipol2* is by far the most time-consuming algorithm since all types occur more frequently than in both other variants. However, whether to prefer *GMDIpol2* or *WIpol2* is a more difficult decision. To pinpoint one of them we need to add a different metric of choice, namely code and data size depicted in Figure 4.9. *GMDIpol2* is in favor with respect to *Ipol2* and *WIpol2*. We see the price paid for the acceleration achieved in *WIpol2* by transforming the data tables. While code size is kept stable, data size increased by more than a factor of 10. If the additional storage requirements are not too costly for the embedded system, *WIpol2* seems to be the fastest algorithm. To confirm this hypothesis we group all assembly instructions into the two categories 'mul, div, rem' and others. The first group contains all expensive, i.e. time consuming operations. Now let x be the ratio of the runtime cost for instruction types mul, div, rem to the runtime for other instructions. Figure 4.11 can now be expressed as an equation for the determination of the fastest algorithm. The equation

$$71709 + 10824x = 168927 + 7623x$$

is satisfied for x = 97218 / 3201 ≈ 30. For this x runtimes for both algorithms equate. Hence we can conclude: if the operations mul, div, rem take about 30 times longer than the remaining operations on a micro-controller, *GMDIpol2* is the faster algorithm, otherwise, *WIpol2* is faster. For example, on a Sparc IPC, the ratio is approximately 1/25, i.e. *WIpol2* is faster. On an 8086 processor, however, it is exactly the other way round.

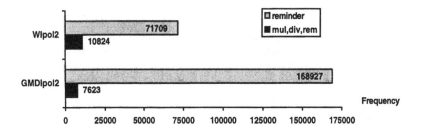

Figure 4.11. Ratio of instruction types for two different implementations of two dimensional interpolation

This calculation very nicely proves the interdependencies between algorithm optimization and hardware selection. None of the two decisions, neither for a given hardware nor for a software, can be made independently. Since we fixed IPC as the target processor, *WIpol2* is the recommended software choice. Paladin proves to be an indispensable tool for this decision.

4.4.6 Cache Estimation

Almost all processors support data and/or code caches today. This section deduces a rough algorithm which can accuracy be expected from runtime determination if the considered system includes a cache.

For 150003 *GMDIpol2* calls on IPC, Paladin monitored 3.9 million   load instructions, 0 write instructions, 25.2 million total instructions using 29.2 million cycles. This data is now used to estimate the maximum run time delay under the worst case assumption that there we have a cache miss for every load instruction. Execution time without cache misses amounts to 29196670 clocks / 150003 calls ≈ 200 clocks per call. A cache miss costs 4 additional clocks and loads a 64 byte cache line which contains 16 instructions. Under worst case conditions for the instruction cache, this are roughly ≈ 1.58 million instruction cache misses for 2.5 million instructions and therefore 6.3 million cycles per 150003 calls. A penalty of 42 cycles per call results. For data cache calculation, the number of the load and write instructions is 3942845, this requires ≈ 15.7 million clocks of data cache errors, i.e. 105 clocks per call. It is assumed here that all data accesses cause an error and not just every 16-th access as in the case of the instruction cache. This estimations show an execution time variation from 200 to 347 clocks per call of *GMDIpol2*. However, it has to be noted that this estimation span is too large an interval due to the extremely unfavorable data cache estimation. A maximum uncertainty of 20 % would be realistic in case of normal program execution.

4.4.7 Optimization of the Specification

Recall from the specification section that in the C program injection calculation is activated every msec. But a limit rpm of e.g. 5000 with 4 cylinders makes injection necessary about every 3 msec. Thus the fuel quantity is computed too often. The intermediate computations do not influence the performance of the motor. They do have an impact on the stability of the control algorithm, though, which is a proportional-integral controller (PI). Subsequent computations are modified by values of internally stored variables. In the following experiment we vary the two calling frequencies f_1 and/or f_2 to function *Steuergeraet* and PI controller (included in *Steuergeraet*) respectively.

Table 4.6 shows the impacts of lowering f_1 on the injection quantity driving along the standard driving curve, Figure 4.4. In particular from 2 to 30 s when the motor is idle, the transient period of the control is extended with reduced call frequency f_1. It gets completely out of clock at 63 Hz and crashes. Also the non-steady state oscillation is increased. In Table 4.7, f_1 = 1Khz was maintained and f_2 was reduced. Clearly the observed distortions are less pronounced. A combination of f_1=500Hz with f_2=63Hz was compared point-wise to the original injection output in Table 4.7, last picture, dotted line. It appears to be a tolerable controller behavior. From this reduced frequency computations, the following time savings resulted. With original frequencies f_1=f_2=1Khz the two-dimensional interpolation Ipol2 is activated 150003 times and the PI controller 50001 times, with f_1=500 Hz and f_2=63Hz the two-dimensional interpolation is activated only 9378 times and the PI controller only 3126 times. This accelerates the computation of the injection quantity by a factor of 3. Compared to the original floating point version, we even gain a factor of 4.9.

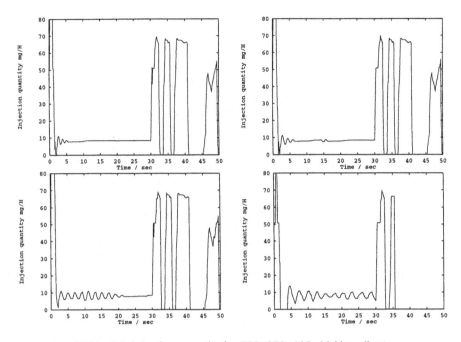

Table 4.6. Injection quantity for 500, 250, 125, 63 Hz call rate

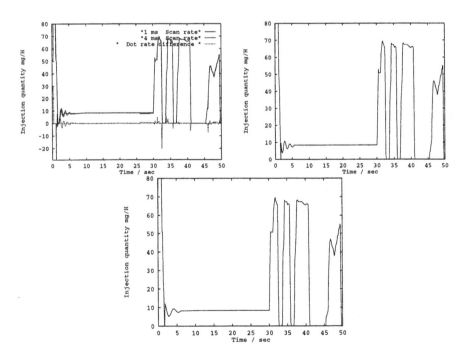

Table 4.7. Injection for 250, 63 Hz call rate PI controller and point-wise difference

4.4.8 Summary

This section demonstrated CASTLE usage working on the EDC code. Separation of the example program into a simulation and controller part and the elimination of unnecessary floating calculation involved manual intervention. A top-down analysis methodology was applied.

A thorough program analysis unveiled performance hot spots with Procov. Then a performance prediction using Kukl takes place. It includes some worst case scenarios and indicates a suitable controller. These global observations are doubly verified, first locally by concentrating on the most demanding function *Steuergeraet* then by compiling to target assembly. The following detailed code analysis produces an additional criterion which hardware to chose for the given specification. It relates instruction distribution to instruction timings and confirms the initial choice. Subsequent code optimizations quantify space/time trade-offs in detail. Processor architecture is accounted for and briefly estimates worst case execution time in presence of caches. The attempt to modify the rates given in the initial specification to accelerated the execution speed even further concludes the investigation. As expected, biggest gains come from requirement variations, while elimination of floats or algorithmic transformations only gain some 25%. It is furthermore imperative to have quality metrics included into the specification to evaluate transformations in accordance with the embedded system and its environment.

4.5 Conclusions

The EDC is a system which is embedded in a technical process, namely a vehicle. Its timing depends on this process. The tasks to be performed by an embedded system within the overall system has to be described in order to implement the subsystem adequately. The functional requirements, including the timing constraints, must be known. The more automated the embedded system should be produced the more formalized the tasks and requirements should be. Above all the verification of temporal security requirements demands such a formalization which improves the visibility of target conflicts. The C program was not suitable for verifying temporal security requirements which was due to the close coupling of the simulation of the technical process and the embedded control.

Nevertheless it is of utmost importance for the analysis and to answer typical questions arising in embedded system design, which are likely to arise in many other design examples. These are as follows:

- does a standard micro-controller met the constraints and run the implementation fast enough? What metrics can be used to make a justified choice?

- where are performance hot spots, how can they be optimized and what will be the resulting acceleration?

- how do measurement, computing and control accuracy influence the control behavior?

- how are technical process and the computing processes of the embedded system coupled. What impact might a change have on the stability of the whole system consisting of both?

All these questions have been examined on the EDC example and by the help of the CASTLE tool suite. The examination and its results have been described. The most time-consuming parts of the control have been identified and different algorithmic variants for acceleration have been examined. CASTLE shows that from a set of candidate processors i80386SX, 25 Mhz, i80486DX, 33 Mhz and LSI 64801, 25 Mhz only the last two are fast enough, though the i80386SX is on the verge. The selection of the implementation with minimum CPU time depends on the controller to be used. This interdependence was demonstrated and quantified by means of CASTLE (in section 4.4.5).

The most time-consuming part of the present system description is the two-dimensional interpolation. The greatest program acceleration is obtained by the reduction of interpolations calls which in turn is achieved by suitable changes to the call rate of the EDC.

Though the present example did not enable a conclusive examination, it provided extremely valuable information on the co-design of an industrial application.

References

[ETAS97] ETAS GmbH, Development for Embedded Control Systems, *Whitepaper Concepts and Technology*, www.etas.de, D-70469 Stuttgart, 1997.

[Int92] Intel Corporation, *Intel486 Microprocessor Family Programmer's Reference Manual*, Santa Clara, California, 1992.

[Kir88] M. Kirschner, *Elektronische Dieselregelung*, Vortrag an der Technischen Akademie Esslingen, 20 pages, 2.2.1988.

[Kor84] W. D. Körner, G. J. Fränkle, *Elektronische Dieselregelung EDR für Nutzfahrzeug-Motoren, VDI-Berichte Nr. 515*, 8 pages, 1984.

[Plo97] P. G. Plöger, H.T. Vierhaus, Retargierbare Laufzeitabschätzungen, *8. EIS Workshop*, University of Hamburg, 8.-9. April 1997.

[Rei92] R. Reißig, Implementierung und Test von Moduln zur Simulation einer elektronischen Dieselregelung in der Programmiersprache C, *Ingenieurarbeit an der Ingenieurschule für Elektronik und Informationsverarbeitung Görlitz*, 29.5.1992.

[The96] M. Theißinger, P. G. Plöger, H. Veit, W. Göhring, G. Pfeiffer, E. Moser *Untersuchung zum Codesign einer Dieseleinspritzsteuerung mit CASTLE, GMD-Studien Nr. 306*, ISBN 3-88457-306-3, 1996.

[Wol94] W. H. Wolf, *Hardware-Software Co-Design of Embedded Systems*, Proceedings of the IEEE, Vol. 82, No. 7, July 1994.

5 AUTOMATIC FORMAL DERIVATION APPLIED TO HIGH-LEVEL SYNTHESIS

José Manuel Mendías
Román Hermida

5.1 Introduction

Since the early 80's, when first high-level synthesis (HLS) algorithms appeared, behavioral synthesis tools have had to evolve quickly. From day to day new fields of application are being discovered that require to review the old techniques. Continuously, the market is requiring new algorithms able to perform more efficient search of solutions. Everyday more abstract specification methods are required, which enlarge the semantic gap between specification and circuit. And all this development has got a price to be paid: the complexity of algorithms grows and the supporting data structures become more sophisticated. As a consequence, the bugs in the tools (or even in the algorithms) proliferate. The effect is that reliance in synthesis tools decreases and nowadays no sensible designer takes the risk to accept a circuit automatically generated without a later validation step; either using simulation or verification techniques. This means, in practice, that the well-known paradigm of *correctness by construction*, widely publicized formerly, has to be handled very carefully.

To address this problem the so-called formal or transformational synthesis systems appeared recently. Their aim is to perform all the design steps within a purely mathematical framework, where the synthesis process itself becomes the proof of soundness of the implementation. There are three common characteristics of these systems: i) they use a single mathematical formalism that is versatile enough as to represent the specification, the implementation, and the set of possible intermediate states of a design process; ii) they synthesize by sequentially applying a set of behavior preserving transformations, which have been proved to be correct;

J.C. López et al. (eds.), Advanced Techniques for Embedded Systems Design and Test, 103-123.
© 1998 *Kluwer Academic Publishers.*

iii) they are not automatic: although any transformation can be done automatically, the sequence of transformations is decided by the designer. From a theoretical point of view, the designs generated by these systems are necessarily correct by construction, given that the transformations are sound. However, from a practical point of view, the reliability of the tool depends both on the complexity of the mathematical support and the reliability of the implementation of the kernel of transformations.

Although a good survey on formal systems can be found in [Kuma96], some systems can be highlighted. The DDD system [JoBo91], which being based on a simple functional language, is focused to manually transform datapaths under sequential control. It uses the so-called factorizations [John89] as basic transformations to modify the design structure by applying distributivity among functions. Another interesting system is T-Ruby [ShRa95] which works with very abstract specifications described in terms of input/output relations at both logic and RT-level [JoSh90]. It allows to obtain a great deal of implementations with systolic or iterative style. In the commercial side, the LAMBDA/DIALOG [FiFM91] tool should be mentioned as an example of the application of higher order theorem provers to design problems. And more recently, we could cite HASH [BlEi97] that is specialized in the verification of schedules of linear graphs, by using as a kernel a modified version of the higher order theorem prover HOL. There are two important aspects to be considered in HASH. Firstly, it does not use a standard implementation of HOL, because in standard implementations the complexity of the proof grows exponentially with the length of the critical path. Secondly, the proposed representation, unfortunately, does not admit feedback loops, which means a severe limitation on input graphs and, even worse, the inability to represent real circuits, that must be fed back to allow component reuse.

In this chapter we will present the main features of a research project that led to the design of the formal tool FRESH (FRom Equations to Hardware) that covers the whole HLS process. Like previous systems, the tool is not self-contained (it does not make any design decisions), but besides the possibility of being operated by a designer, it can be driven by a conventional HSL tool, thus creating a framework where the correctness of the design can be ensured automatically. The main advantages of this system are easiness, reliability, applicability and efficiency. Easiness means not having to modify conventional tools to adapt them to the mathematical formalism; they simply must deliver the results of their search: what cycle each operation has been scheduled in, what operations share the same module, etc. Reliability means selecting a kernel of few simple transformations (which minimizes the number of error sources) and adopting a declarative representation with first-order formal semantics (which simplifies both designer interpretation and tool processing). Applicability is obtained, differing from other systems, neither by constraining the kind of accepted behaviors nor the kind of reachable designs. Efficiency is obtained by specializing the system in order to perform complete formal synthesis processes (from specs to datapath+controller) with quadratic complexity.

The chapter will show the basic ideas of our formal approach taking into account the intended audience. Thus, it will try to avoid intricate mathematical details, while keeping the integrity of the message. However, for those readers wishing to go further with the formal support, a specific appendix has been included. In section 5.2 we will present the behavior representation, discuss how to obtain such representation, and present a set of operators, which will be used in section 5.3 to formalize the HLS stages. In section 5.4 we will show the kernel of behavior preserving transformations which allow us to perform formal synthesis. Sections 5.5 and 5.6 are devoted to study the transformational algorithm and its complexity. Some conclusions and future work are summarized in section 5.7.

5.2 System Representation

To derive a circuit implementation from a given spec, it is necessary that both, implementation and spec, as well as, all the intermediate states (e.g. partially scheduled or partially allocated graphs) share a single representation formalism. The aim of this section is to present that common representation.

It is well-known that a combinational circuit, alike a linear data-flow graph, can be represented by a set of equations describing their structure. Every equation of this set defines how to compute an internal signal, or an output port, in terms of a composition of operators whose arguments are other signals and/or input ports. We suppose that there is a sufficient collection of operators to perform the usual arithmetic and selection operations. We call them non-temporal operators. If we, additionally, want the set of equations to be able to describe fedback sequential circuits or iterative computations, we must add a temporal operator to represent delay. This operator, that we call *fby* (followed by), must appear in every computation loop. We call this sort of representation *equational spec*, and in figure 5.1.a the equational spec of a 2nd order recursive filter is shown.

An equational spec, besides the schematic interpretation mentioned above, has a more useful semantics in terms of streams. A stream [Delg87] is a sequence (infinite for our purpose) of values belonging to certain domain in correspondence with those ones carried by a signal along time. If a signal denotes a stream, every non-temporal operator will denote a pointwise stream function; every constant will denote a stream with a single value infinitely repeated; and *fby* will be the operator that inserts an atomic value in the head of a stream.

5.2.1 Getting Equational Specifications

Getting equational specs from procedural ones (i.e. VHDL) is not difficult. It is enough to compile the source code and describe the resulting graph in terms of a set of mutually recursive equations. However, in many applications it is easier to bypass the elaboration of procedural code, since equational specs can be directly obtained from alternative specification styles that are more natural for the designer. Thus, to describe block diagrams equationally (figure 5.1.b) it is enough to use the structural interpretation of our representation: for each wire a name must be provided first, and then its definition is derived from the functionality of the block that is driving the wire. For combinational blocks the equation will be of the form *OutputSignal = Operator(ListOfInputSignals)*, and for delayers we will have *OutputSignal = InitialValue fby InputSignal.* A complex or hierarchical block will be represented by an equation subset.

On the other hand, to specify equationally those behaviors using explicit temporal indexing (figure 5.1.c) it is enough to use the interpretation as streams of our representation. The complete history of values carried by each variable $x(t)$, with t ranging in N_+ (i.e. $\langle x(0), x(1)\ x(2), ... \rangle$), is a stream, and so, can be represented in our formalism simply by x. Correspondingly, the complete history of values carried by the variable $x(t-1)$, with t ranging in N_+ and $x(t')=0$ whenever $t'<0$ (i.e. $\langle 0, x(0), x(1)\ x(2), ... \rangle$) will be represented by $0\ fby\ x$. Obviously, $x(t-2)$ can be expressed by $0\ fby\ 0\ fby\ x$, and so on.

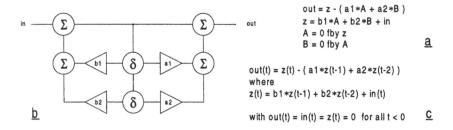

out = z - (a1*A + a2*B)
z = b1*A + b2*B + in
A = 0 fby z
B = 0 fby A   <u>a</u>

out(t) = z(t) - (a1*z(t-1) + a2*z(t-2))
where
z(t) = b1*z(t-1) + b2*z(t-2) + in(t)

with out(t) = in(t) = z(t) = 0 for all t < 0   <u>c</u>

<u>b</u>

Figure 5.1. Equational (a), structural (b) and temporal (c) specifications

5.2.2 A Set of Temporal Operators

With a sufficient amount of non-temporal operators (combinational) plus the *fby*, it is possible to describe any synchronous circuit and a large number of computations [WaAs85]. However, if we want to represent all intermediate stages between algorithms and circuits, it is necessary to find some extra temporal operators which can express the key idea of any HLS process: the time multiplexed use of hardware resources. Two main factors are crucial to support that temporal multiplexing: i) the internal register transfer frequency should be greater than the input data sampling, and ii) the hardware resources should be reused to perform several operations of the data-flow graph in different instants. Thus, we will define the operators << *(replicate)* and >> *(sample)* to express the frequency ratio between two signals, and the operator || *(interleave)* to express multiplexed use of resources. In addition we will use the symbol # *(wildcard)* to represent those values computed but not stored in certain cycle (i.e. calculated by a resource which no operation has been assigned to). And, finally, we will introduce the *next* operator as the inverse of *fby*.

Let x be a signal, $x \ll n$ is the signal that runs at a frequency n times higher than x, transporting the same sequence of values as x, but repeating each one n times. $n \gg x$ is the signal that runs at a frequency n times lower than x, transporting 1 out of every n values carried by x. The first operator models fast devices reading slow signals: the same value is read several times. The second operator models the opposite, slow devices reading fast signals: some values are read and some other are lost. Let x_j be a collection of k signals, then $(x_1 \Vert ... \Vert x_k)$ is the signal that periodically interleaves the values carried by its k arguments, picking up every instant one value and discarding the remaining k-1 ones. It models what happens after a k-cycle scheduling: a signal can transport, in different cycles, values coming from a maximum of k different sources. To denote a generic data source we use the constant signal #. Finally, if x is a signal, *next x* will be the signal which carries all x values except the first one. It is non causal and so non implementable. However it has a fundamental role to represent partially scheduled graphs because next allows an scheduled operation to express the possible cycles in which its predecessors should be scheduled (even though they are not yet). As an example, let $index = \langle 1, 2, 3, 4, ... \rangle$ be an stream, then:

$$(index \ll 2) = \langle 1, 1, 2, 2, 3, 3, 4, ... \rangle$$
$$(2 \gg index) = \langle 2, 4, 6, 8, 10, 12, 14, ... \rangle$$
$$(\# \Vert index) = \langle \#, 2, \#, 4, \#, 6, \#, ... \rangle$$
$$(next\ index) = \langle 2, 3, 4, 5, 6, 7, 8, ... \rangle$$

5.3 High Level Stages Formulation

Once we have got the minimum operator set to express any partial design, we are going to present a set of properties that they fulfill. With these properties (which have been previously proved) we intend to formalize mathematically every stage of a HLS process and to behold the task of designing as a simple first order equational calculus process. The properties will be briefly shown as families of equations, that we can classify in 5 types:

i) *Inverse operator:* there are two of them, IFBY states that next is the inverse operator of fby, and IREP states that sample is replicated.

ii) *Temporal distributivity:* there are five families, DFBY, DNEXT, DSAM, DREP y DINT, they state, respectively, the distributivity of temporal operators with respect to non temporal ones.

iii) *Identity elements:* there are five, NFBY, NNEXT, NSAM, NREP y NINT, they state that any constant signal is not affected by temporal operators.

iv) High level synthesis: there are five families that are detailed below:

Temporal multiplexing theorems (TMT), that allows to assign a cycle to compute an operation tagging remaining cycles with *wildcards* for later reuse. The support idea is: if the values transported by a signal are read with less frequency than they are produced, a lot of them will not have any effect on the final computation and will be able to be replaced by any others. As these values are calculated by certain fast operator, all those cycles in which a value is produced but not read will be able to be used to perform any other useful operation. This happens whenever the operator receives its arguments in the proper moment and the valid result is stored during enough cycles to arrive on time to other operators. As an example, we are going to show how the temporal operators are used to formulate equationally this idea (notice the formula has got as parameters k and m):

$$k \gg x = k \gg (\# \textit{ fby })^m (\# \| \overset{k-m-1}{\ldots} \| \# \| next^m (x) \| \# \| \overset{m}{\ldots} \| \#)$$

As it can be seen, we use the *sample* operator to express the ratio k between external and internal frequency, and the *wildcard* to express the carried but not read values. Through the interleave operator with k arguments which has got $next^m(x)$ as $(k-m)$-th argument, we express the assignment of x operation to cycle $k-m$ within a k cycles scheduling. By means of *next* and *fby* operators we state the temporal requirements of that operation with regard to the rest. When all the operations are scheduled the *fby* operators affecting each operation and the *next* operators affecting its successors will have to cancel in pairs by using the equation *IFBY*.

Architectural delayers replacement theorems (ADRET), that allows to use architectural registers to store auxiliary values in those cycles in which the registers cannot be observed. An architectural delayer stores values between two or more initiations of the algorithm, thus it has the same load frequency as the data sampling. Nevertheless, if we perform HLS a new kind of auxiliary delayers appear. They store temporal values within an initiation and have a higher load frequency. The idea is to turn architectural delayers, whose reading can be scheduled with the previous theorem, into chains of auxiliary delayers.

$$(\# \| \overset{k-m-1}{\ldots} \| \# \| next^m ((y \textit{ fby } (k \gg x)) \ll k) \| \# \| \overset{m}{\ldots} \| \#) =$$

$$= (\# \textit{ fby })^{k-m-1} y \textit{ fby } (\# \textit{ fby })^m (\# \| \overset{k-m-1}{\ldots} \| \# \| next^m (x) \| \# \| \overset{m}{\ldots} \| \#)$$

Memorization theorems (MEMT), that allows to replace chains of delayers by a feedback one. The support idea is rather simple, when a value is computed in certain cycle $k-m$ but it has to be delayed $n+1$ cycles to be used, it possible to replace the $n+1$ delayers by just 1, if this delayer is fed back during n cycles. We have two cases, when the values are computed and consumed within the same algorithm initiation:

$$y \; fby \; (\,\# \, fby \,)^n \; (\,\# \| \overset{k-m-1}{\ldots} \; \| \# \| x \| \# \| \overset{m}{\ldots} \| \#\,) \approx$$

$$\approx fix(\; \lambda z.(\; y \; fby \; (\,\# \| \overset{k-m-1}{\ldots} \; \| \# \| x \| z \| \overset{n}{\ldots} \| z \| \# \| \overset{m-n}{\ldots} \; \| \#\,)\,)\,)$$

and when the values are computed in an initiation but consumed in next one:

$$y \; fby \; (\,\# \, fby \,)^n \; (\,\# \| \overset{k-m-1}{\ldots} \; \| \# \| x \| \# \| \overset{m}{\ldots} \| \#\,) \approx$$

$$\approx fix(\; \lambda z.(\; y \; fby \; (\, z \| \overset{n-m}{\ldots} \| z \| \# \| \overset{k-n-1}{\ldots} \; \| \# \| x \| z \| \overset{m}{\ldots} \| z\,)\,)\,)$$

The *fix* operator [Stoy77] allows to express anonymous recursivity, and $\approx$ indicates that both streams never transmit different values (although one of them can transmit a wildcard and the other one a ordinary value).

Decomposition theorems (DET), that allows to separate the different RT-level actions included in a high-level operation (i.e. operand selection, computation, result storage and data transfer) in order to separately reuse the different hardware modules involved.

$$(\, x_1 \| \overset{i-1}{\ldots} \| x_i \| \overset{k-i}{\ldots} \| x_k \,) =$$

$$= (\, x_1 \| \overset{i-1}{\ldots} \| x_{i-1} \| (\, \# \| \overset{i-1}{\ldots} \| \# \| x_i \| \# \| \overset{k-i}{\ldots} \| \#\,) \| x_{i+1} \| \overset{k-i}{\ldots} \| x_k \,)$$

Input anticipation theorems (INAT): this theorem states that a value read in a slow input port is the same in any cycle, so it can be foreseen.

$$(\, x_1 \| \overset{k-m-1}{\ldots} \; \| x_{k-m-1} \| next^m (x \ll k) \| x_{k-m+1} \| \overset{m}{\ldots} \| x_k \,) =$$

$$= (\, x_1 \| \overset{k-m-1}{\ldots} \; \| x_{k-m-1} \| x \ll k \| x_{k-m+1} \| \overset{m}{\ldots} \| x_k \,)$$

Multiplexer implementation (MUXI): this property states the equivalence of an *interleave* operator whose k arguments are n, maybe repeated, data sources ($n \leq k$), and a multiplexer n to 1, whose arguments are non repeated data sources and control is generated by an *interleave* operator with k constant arguments (generating the source selection pattern).

$$(\, x_{j1} \| \overset{k}{\ldots} \| x_{jk} \,) = mux(\, x_1, \overset{n}{\ldots}, x_n, (\, j1 \| \overset{k}{\ldots} \| jk \,)\,)$$

where $(\, j1, \ldots, jk \,)$ is a permutation of the index set $(\, 1, \overset{n}{\ldots}, n \,)$

v) RT-level implementation theorems: that allows to formalize the operator mapping into hardware modules belonging to certain RT-level libraries [MeHF96].

5.4 A Transformational Design Kernel

After the formalization of the different HLS stages as equations, we need a set of manipulation rules that allow to apply them properly to transform an equational spec into another one with the same behavior, but different cost-performance. All the rules have been proved to be correct and constitute, as a whole, the only computation mechanism allowed in our formal synthesis system. Given that this kernel is small and simple, we have been able to reduce to a minimum the risk of programming errors. We can distinguish two types of rules: structural, that simply modify the equations layout, and behavioral, that modify the way the equational spec performs the computations.

The summary of the set of structural rules is:

i) *Substitution:* allows to replace any occurrence of a signal by its definition.

ii) *Rename:* allows to change the name of a signal.

iii) *Expansion:* allows to replace any subterm of a definition by a new signal if the signal is defined as the subterm to be replaced.

iv) *Elimination:* allows to remove any definition not used by any other.

v) *Cleaning*: allows to remove redundant definitions.

The summary of the set of behavioral rules is:

i) *Replacement*: allows to replace any *wildcard* by any other term denoting whatever concept (to be understood in terms of reuse).

ii) *Rewriting left-right*: transforms a definition applying a universal first order formula from left to right.

iii) *Rewriting right-left*: transforms a definition applying a universal first order formula from right to left.

5.4.1 An Example of Interactive Formal Design

As an example consider the following formulae where all variables are universally quantified:

[a] $(x_1\,fby\,y_1)*(x_2\,fby\,y_2) = (x_1*x_2)\,fby\,(y_1*y_2)$

[b] $a2\,fby\,a2 = a2$

[c] $x*0 = 0$

Formula [a] expresses the distributivity of the *fby* operator in relation to arithmetical multiplication (DFBY), which in terms of hardware must be interpreted as retiming. Formula [b], expresses that a constant stream is not affected by the *fby* operator (IFBY), and formula [c], states an obvious arithmetical property over streams.

Starting from the equational specification shown in figure 5.1.a, we use the substitution rule to replace all the A and B occurrences by their definitions. Definitions now becoming dead code may be removed by using the elimination rule. After that using three times the rules of rewriting, it is possible to obtain a retimed definition of signal out, by applying on it formulae b, a and c.

$$z - (a1*(0 \; fby \; z) + a2*(0 \; fby \; 0 \; fby \; z)))$$

$$=_{[b]} z - (a1*(0 \; fby \; z) + (a2 \; fby \; a2)*(0 \; fby \; 0 \; fby \; z)))$$

$$=_{[a]} z - (a1*(0 \; fby \; z) + (a2*0) \; fby \; (a2*(0 \; fby \; z)))$$

$$=_{[c]} z - (a1*(0 \; fby \; z) + 0 \; fby \; (a2*(0 \; fby \; z)))$$

In the same way (using similar equations and applying seven times the rules of rewriting we can retime the definition of z, obtaining the following retimed equational specification:

```
{
    out = z - (a1*(0 fby z) + 0 fby (a2*(0 fby z)))
    z = 0 fby (b1*z + b2*(0 fby z)) + in
}
```

Then in order to reach a more clear description, we can expand every occurrence of the subterm 0 fby z, clean the specification and rename the remaining definition as A, resulting:

```
{
    out = z - ( a1*A + 0 fby (a2*A) )
    z = 0 fby (b1*z + b2*A) + in
    A = 0 fby z
}
```

From the designer's point of view we have formally reduced the critical path of the computation from 4 to 3 operators. Later on we will exploit that change of operations' mobility to improve hardware reuse.

5.5 Automating Formal Design: A Derivation Algorithm to Perform HLS

In contrast with the previous example, where we manually drove the rule application sequence, in this section we are going to present an algorithm which describes how the rules must be sequenced to perform HLS. Remember that this algorithm does not perform design space exploration, but it applies the decisions already made by a conventional tool, in order to prove they are mathematically correct. From now on, we assume that all the information about the synthesis decisions is available: what cycle each node is scheduled in, what operations share the same module, how registers are reused, what the signal order at multiplexer inputs is, etc. This information, along with the equational spec of the circuit, are the inputs of the algorithm (figure 5.2). The output is either another equational spec representing the designed circuit, or a report about the erroneous design decisions made by the tool. The key idea of the algorithm to obtain a *k*-cycle circuit is described next. The original spec is understood as a single cycle circuit having all the operators chained. Our aim is to transform this circuit into another one that, keeping the external data sampling frequency, works internally *k* times faster, may have non-chained operators, and may reuse both registers and operators. We will do this gradually, starting from data sources until the data drains are reached, we will increase the frequency of each operator and schedule it in a cycle. The data sources for any algorithm initiation will be the input ports, the constant and the outputs of the architectural delayers. The data drains will be the output ports and the inputs of the architectural delayers.

inputs: spec, design decisions
outputs: circuit, were decisions correct?

    normalization
    source multiplexing
    architectural delays scheduling
    for 1 to critical path length do begin
        sample export
        sample extract
        operation scheduling
    end
    next elimination
    scheduling correctness check
    action decomposition
    delayers feedback
    module reuse
    allocation correctness check
    multiplexer synthesis
    control reuse

Figure 5.2. Algorithm scheme

Following the scheme shown in figure 5.2, we start by normalizing the spec to detect errors (such as combinational feedbacks) and to simplify subsequent processing. The canonical spec obtained has no dead code, no redundancies and each equation within it has just one operator. This phase uses several times the rules of *expansion*, *cleaning* and *elimination*.

After that, we increase sources read frequency applying *IREP* equation, the architectural delayers update cycle is scheduled applying *TMT* and *ADRET*.

Then, the *sample* operator is spread among the equations in two phases. The export phase moves the *sample* operator from the output of a just scheduled operation to the inputs of its successors, the *expansion* and *substitution* rules are used. The extract phase uses the equation *DREP* to turn a slow operator with *sampled* inputs into a *sampled* fast operator with non *sampled* inputs.

```
{
    out = 4 >> t33
    t24 = # fby ( t23 + (in<<4) || t24 || t24 || # )
    t33 = # fby ( # || # || t32 - t31 || # )
    t23 = 0 fby ( # || # || # || t27 + t21 )
    t28 = 0 fby ( t28 || # || # || a2*t15 )
    t31 = # fby ( # || t29 + t28 || # || # )
    t32 = # fby ( t23 + t22 || t32 || # || # )
    t21 = # fby ( # || b2*t15 || t21 || # )
    t27 = # fby ( # || # || b1*t32 || # )
    t29 = # fby ( a1*t15 || # || # || # )
    t15 = 0 fby ( t15 || t15 || t15 || t24 )
}
```

Figure 5.3. Four cycles scheduling without reuse

As soon as an operator works at higher frequency we use the equation *TMT* to *schedule* it in the corresponding cycle. Once the *sample* operator reaches the drains, the scheduling is finished and the data dependencies must be checked. To do this, we try to eliminate all the *next* operators applying the *IFBY* equation on the internal transfers and the *INAT* on the input ports. If this is achieved then the scheduling is correct, otherwise incorrect.

Once the scheduling is done and checked we apply *DET* to decompose every equation that shows the assignment of an operation to a cycle into two another ones: one representing the computation itself and other representing the storage of intermediate values. The chains of delayers are replaced by fedback ones using the equation *MEMT*.

Next we check the correctness of the reuses, to do this it is enough to substitute some *wildcards* and to eliminate the redundancies that spring from these substitutions (the *replacement* and *cleaning* rules are used). If the number of final operators is equal to the number fixed by the allocation stage, it is correct otherwise incorrect.

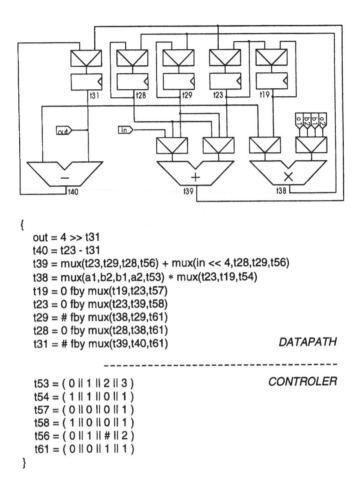

```
{
    out = 4 >> t31
    t40 = t23 - t31
    t39 = mux(t23,t29,t28,t56) + mux(in << 4,t28,t29,t56)
    t38 = mux(a1,b2,b1,a2,t53) * mux(t23,t19,t54)
    t19 = 0 fby mux(t19,t23,t57)
    t23 = 0 fby mux(t23,t39,t58)
    t29 = # fby mux(t38,t29,t61)
    t28 = 0 fby mux(t28,t38,t61)
    t31 = # fby mux(t39,t40,t61)                    DATAPATH

    ------------------------------------
    t53 = ( 0 || 1 || 2 || 3 )                      CONTROLER
    t54 = ( 1 || 1 || 0 || 1 )
    t57 = ( 0 || 0 || 0 || 1 )
    t58 = ( 1 || 0 || 0 || 1 )
    t56 = ( 0 || 1 || # || 2 )
    t61 = ( 0 || 0 || 1 || 1 )
}
```

Figure 5.4. RT-level equational specification of a fully designed circuit

Knowing the correspondence between multiplexers input order and data transfers and using the *MUXI* theorem, it is possible to replace any interleave operator by a multiplexer and a constant pattern control signal.

Finally, it is even possible to reuse control lines replacing *wildcards* again and cancelling redundancies.

Figure 5.3 shows an equational spec of the previously retimed circuit after a 4 cycle scheduling without reuse. Figure 5.4 shows this circuit fully designed (just with RT mapping to go) and a drawing of the datapath.

5.6 Experimental Results

From a theoretical analysis of the algorithm we could see that its complexity depends on 2 factors: the number of nodes in the graph and the number of cycles in the scheduling. The critical path length, that appears as a parameter in the main loop, has no effect due to the fact that the loop aim is to allow that the nodes are scheduled in a suitable order (not necessary unique), from sources to drains. This study concludes with the fact that the temporal complexity of the algorithm is quadratic respect to the number of nodes and linear respect to the number of cycles.

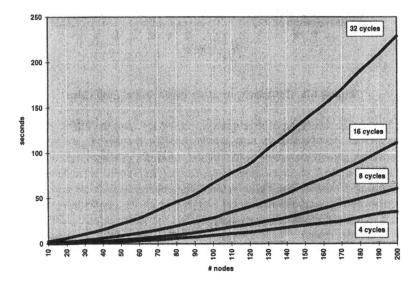

Figure 5.5. Complexity analysis, effect of the schedule length

In order to experimentally contrast the theoretical predictions, we have developed a prototype of the formal system using the programming language PROLOG. Though it is not a language oriented to obtain high performance programs, we have chosen it because powerful symbolic processing applications can be created easily with it. Nevertheless it must be highlighted that our study is not about performance but about complexity. We want just to know how different factors affect the growth of computation time. The graphs used in all experiences have been collections of 2nd order recursive filters in parallel. Each filter (as can be seen in figure 5.1.a) has an input, an output, 4 constants, 10 operators and a critical path of 4 nodes. The operators are 2 architectural delayers, 4 multipliers, 3 adders and 1 subtractor.

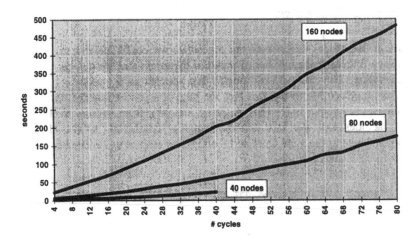

Figure 5.6. Complexity analysis, effect of the graph size

We have made two kind of experiments, one to measure the effect of the graph size and another the effect of the schedule length. The first (see figure 5.5) explores how the computation time grows as the number of nodes increases, keeping constant the schedule length. It has been done by increasing the number of nodes from 10 to 200 in steps of 10 nodes. The figure presents the obtained results for 4, 8, 16 and 32 cycles respectively. Regression analysis shows that all the curves have a quadratic growth. The second group of experiences studies the behavior of computation time against schedule length. These experiences have been done by increasing the number of cycles from 4 to 80 in steps of 4 cycles, for a fixed number of nodes. Results obtained for 40, 80 and 160 nodes are presented in figure 5.6. Regression analysis shows that all the curves verify a quasi-linear behavior.

5.7 Conclusions and Future Work

The main conclusion that can be pointed out with this chapter is that the cooperation of conventional synthesis techniques with formal design ones can generate high quality products that are effective as well as reliable. Within this framework, the contribution that our research brings out is that this cooperation has been developed in an efficient way, that is to say, our aim was to obtain, keeping the mathematical rigor, a system of high practical usefulness (two concepts that may sometimes look like opposite). Therefore, differing from other systems that develop a generic and complex formal support that, among some other problems, can be applied to HLS, we have started from HLS problems and developed a suitable mathematical framework for them. That has led to the three main characteristics of the system: i) it does not constrain either the sort of specification, or the sort of reachable design, or the number of HLS stages that it can perform, ii) even though being all the formal aspects very intuitive for a designer, the system can be driven without knowing any of them and thanks to it any conventional tool could direct the formal synthesis, iii) the complexity of the formal process is of the same order than the complexity of the non formal synthesis process and thus it can be included to an automatic synthesis system at no extra complexity.

The immediate future is directed according to four lines of work. The first one, to try to find a new algorithm or even a new set of equations that allow to decrease the complexity of our formal system. The second one, to automate other techniques of design that we presently guide by hand, such as: retiming, functional-pipelining, loop transformation, etc. The third one, to connect the kernel of a proof system with our framework that allows to check the correctness of the great deal of properties that the non-temporal operators fulfill. And the fourth, to explore the possibility of merging in a single framework the formal and non-formal synthesis systems, modifying the conventional design algorithms in order that they use the equational specifications as an internal representation.

5.A Appendix

In section 5.2, we have presented a simplified and intuitive version of our system representation. In it we have avoided quite interesting aspects such as: formal syntax and semantics, data typing and module library spec. In this appendix we address those features following two parallel routes. Firstly, a syntactic one, which provides support for the automation of the approach. And secondly, a semantic one, to state formally what has to be understood by the previous syntax.

5.A.1 Specification of Static Data Types

While external libraries can be supported, our notation has neither built-in types nor operators. So, according to the requirements of the behavior to be described, any needed symbol and its meaning must be defined using *algebraic specifications* [EhMa85].

First of all, we define a *signature* as the tuple (S, Σ) where S is a list of sorts, and Σ is a list of operation symbols, such that Σ describes the domain and range of the operation symbols. Every signature defines a set, $T_\Sigma(X)$, of well formed expressions called terms. If X is a set of free variable symbols not included in Σ, then a term is either a constant (nullary operation), a variable, or any $\sigma(t_1, \ldots, t_n)$ where σ is an operation symbol and t_i are terms. So, a pair of terms defines an equation.

Giving meaning to the signatures, and so being able to interpret terms and equations, requires the definition of mathematical models of the signatures (called Σ-algebras), which is done by associating a support set to each sort and a function to each operation symbol. So, if we define an interpretation function which maps terms to values, it is possible to verify whether the model satisfies an equation.

An *algebraic specification*, SPEC, is composed of a signature an a set of equations. We call model of an algebraic specification, or SPEC-algebra, to any Σ-algebra which satisfies every equation of the specification. Of the many possible models, we will use the so called quotient term algebra, T_{SPEC}, as the semantics of our specification. This algebra is initial in the class of all the SPEC-algebras, and is defined by a set of equivalence classes of ground terms (terms without variables) with the same interpretation in the equational theory defined by the specification.

In figure 5.7, a simple specification of both unsigned and signed positional radix-2 number systems is shown. In it, we declare as operation symbols four unsigned constructors $(1, 0, \Box1, \Box0)$, two signed constructors $(+\Box, -\Box)$ and an overloaded plus operator. The symbol $\Box$ reflects the position of the arguments in relation to an operation symbol. To declare the meaning of the operation symbols we use sets of equations. So, the first two equations state the well-know property of leading zeroes, that in our initial semantics means that different terms such as 010, 0010 or 10 denote the same concept. The third equation establishes the uniqueness of zero. The remaining equations explain the procedure for the addition of both unsigned and signed radix-2 numerals.

specification
  radix2
sorts
  unsigned, signed
operations

0	: → unsigned;
1	: → unsigned;
$\Box$0	: unsigned → unsigned;
$\Box$1	: unsigned → unsigned;
$\Box + \Box$	: unsigned, unsigned → unsigned;
$+ \Box$	: unsigned → signed;
$- \Box$	: unsigned → signed;
$\Box + \Box$	: signed, signed → signed;

equations $\forall$ a, b $\in$ unsigned
  00 = 0;
  01 = 1;
  -0 = +0

$0 + 0 = 0;$
$0 + 1 = 1;$
$1 + 0 = 1;$
$1 + 1 = 10;$
$a0 + b0 = (a+b)0;$
$a0 + b1 = (a+b)1;$
$a1 + b0 = (a+b)1;$
$a1 + b1 = ((a+b)+1)0;$
$(+a) + (+b) = +(a+b);$
$(-a) + (-b) = -(a+b);$
$(+a) + (-b) = +(a-b) \Leftarrow (a>b);$
$(+a) + (-b) = -(b-a) \Leftarrow (a\leq b);$
$(-a) + (+b) = -(a-b) \Leftarrow (a>b);$
$(-a) + (+b)= +(b-a) \Leftarrow (a\leq b);$

Figure 5.7. Algebraic specification of radix-2 positional number system

In figure 5.8, we show how hardware modules can be described, as well, by algebraic specifications. To this, for each module family we want to specify, we use a sort identifier (binAdder), a constructor defining input ports (binAdderOp), and as many observers as output ports the module has (viewResult, viewCarry). Then, using equations, all the functionalities done by the module are declared. A deeper discussion about the suitability of this kind of specifications in behavioral synthesis can be seen in [MeHF96][MeHF97].

From now on, we will assume some basic data types, as well as their most common operators, have been specified: i) the physical types (bit and bit vectors), ii) some abstract data types such as unsigned, signed, or fixed point, and iii) basic RTL libraries in which hardware modules have been specified in algebraic style. We will also assume, that these specifications are correct and have an unique formal semantics in concordance with the informal one normally used by designers. To find a criterion of soundness of algebraic specifications see [EhMa85]. To find algebraic specifications of common numerical data types see [Pada88].

<u>**specification**</u>
  targetLibrary
<u>**sorts**</u>
  binAdder
<u>**operations**</u>
  binAdderOp   : bitVector, bitVector, bit $\rightarrow$ binAdder
  viewResult   : binAdder $\rightarrow$ bitVector
  viewCarry   : binAdder $\rightarrow$ bit
<u>**equations**</u> $\forall$ c $\in$ bit; $\forall$ bv1, bv2 $\in$ bitVector
  add(bv1, bv2, c) = viewResult(binAdderOp(bv1, bv2, b))
  carry(bv1, bv2, c) = viewCarry(binAdderOp(bv1, bv2, b))

Figure 5.8. Algebraic specification of hardware modules

5.A.2 Extension to Express Indifference

Every signature of each algebraic specification and its semantics must be extended with a new element #, called *wildcard*. This will be used to tag every data calculated but not used in later computations, and so it has a clear design meaning: if an operator in certain time slot operates with wildcards, it can be reused to compute any other computation made by any other busy operator.

If Σ is a signature and A is its support algebra, $\Sigma^{\#}$ is the signature Σ plus a *wildcard* per sort and $A^{\#}$ is the algebra derived by extending the universe of A with the *wildcard*, and strictly extending all the functions in A (the notion of strict extension can be found in [Stoy77]).

5.A.3 Extension to Handle Undefinition

Due to the usage of recursive definitions (whose hardware counterpart are feedback networks), we need to provide our algebras with lattice structure by adding another new element ⊥, called *bottom*, to denote undefinition (whose hardware meaning is unstability caused by combinational loops) and a partial ordering, ⊑, called *approximation*. This extension provides us with methods to both detect faulty specifications, and construct reliable simulators.

If A is an algebra, $A_\perp$ is the algebra whose universe is the flat domain defined from A, and its functions are the strict extension of the functions in A (the notion of flat domain can be also found in [Stoy77]).

5.A.4 Temporal Extension

Every atomic object of an algebra, A, is a single item and in general, is unable to describe any dynamic activity. However, if we think of a continuously running operator, we can abstract a mathematical object from this process: the whole sequence of items, or equivalently the application $(N_+ \rightarrow A)$. So, in this phase we define how to extend any data type to another new one, whose objects are infinite streams of temporally ordered single items. To that, instead of defining temporal indexing, which forces designers to do much of the work by themselves and has been pointed out as inadequate [WaAs85], we define just one high level temporal operator: *fby* (followed by), that will allow us to define sequences by recurrence.

Then if SPEC is an algebraic specification with signature Σ_{SPEC}, we define $Lu(\Sigma_{SPEC})$ as the signature, $(\Sigma_{SPEC})^\#$ containing some new polymorphic symbols *fby*, *next*, >>, <<, ‖, and $Lu(T_{SPEC})$ is the only algebra such as:

i) Its universe is the collection of all infinite sequences of elements of the universe of $((T_{SPEC})^\#)_\perp$, that is $(N_+ \rightarrow ((T_{SPEC})^\#)_\perp)$

ii) Any operation symbol $\sigma \in \Sigma_{SPEC}$, denoting the function $(T_{SPEC})_\sigma$, gives birth to the denotation $Lu(T_{SPEC})_\sigma$ of the same symbol in the new signature $Lu(\Sigma_{SPEC})$, as the function which operates pointwise with sequences as follows:

$$Lu(T_{SPEC})_\sigma(x_1, \ldots, x_n) =$$
$$\langle (T_{SPEC})_\sigma(x_1(1), \ldots, x_n(1)), (T_{SPEC})_\sigma(x_1(2), \ldots, x_n(2)), \ldots \rangle$$

iii) The symbols *fby*, *next*, >>, <<, ‖, denote the functions:

$$x_1 \; fby \; x_2 = \langle x_1(1), x_2(1), x_2(2), \ldots \rangle$$
$$next \; x = \langle x(2), x(3), x(4) \ldots \rangle$$
$$x << k = \langle \overset{k}{x(1), \ldots, x(1)}, \overset{k}{x(2), \ldots, x(2)}, x(3) \rangle$$
$$k >> x = \langle x(k), x(2*k), x(3*k), \ldots \rangle$$
$$(\overset{k}{x_1 \, \| \ldots \| \, x_k}) = \langle x_1(1), x_2(2), \ldots, x_k(k), x_1(k+1), x_2(k+2), \ldots \rangle$$

where for all $x \in Lu(T_{SPEC})$, $x(i) \in T_{SPEC}$ references the ith element of the temporal sequence.

5.A.5 System Specification

Now we are ready to define our formalism:

<u>Definition</u>. We call *equational specification* of a digital system to the tuple (SPEC,X,Ins,Outs,φ), where SPEC is an algebraic specification with signature Σ_{SPEC}, X is a family of sets of free variables, such that $\Sigma_{SPEC} \cap Ins = \varnothing$ and Outs are proper subsets of X such that $Ins \cap Outs = \varnothing$, and φ is a function which projects, preserving sorts, elements from X-Ins to the set of terms of the Lu-extended signature:

$$\varphi : (X - Ins) \to T_{Lu(\Sigma_{SPEC})}(X - Outs)$$

<u>Notation</u>. We call respectively *signal*, *input port* and *output port* to each symbol included in X, Ins and Outs. The function φ is called *specification body*, and every pair $(x, \varphi(x))$ appearing on it is called *definition*. If we represent a definition as the equation $x = \varphi(x)$, the specification body can be represented as a set of equations like this:

$$\{ x_1 = t_1, ..., x_n = t_n \} \quad \text{with } t_i \equiv \varphi(x_i)$$

The full equational specification of the second order filter is shown in figure 5.9.

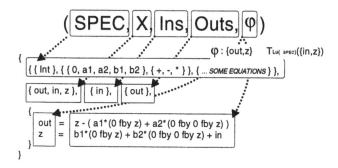

Figure 5.9. Full equational specification

5.A.6 Formal Semantics of an Equational Specification

Our goal now is to relate the set of mutually recursive equations defined by the equational specification with its solution, i.e. the behavior we are attempting to specify.

As a circuit with p input ports and q output ports needs p infinite streams of values, $(N_+ \to ((T_{SPEC})^{\#})_\perp$, to compute another q streams, its behavior will be a function from a p-tuple of signals, to a q-tuple of signals. To obtain formally that behavior we define a semantic function C (circuit), which associates to every specification (syntax) its corresponding behavior (semantics).

$$C : (SPEC, X, Ins, Outs, \varphi) \rightarrow$$

$$\rightarrow (Lu(T_{SPEC})_1 \times \overset{p}{\dots} \times Lu(T_{SPEC})_p \rightarrow Lu(T_{SPEC})_1 \times \overset{q}{\dots} \times Lu(T_{SPEC})_q)$$

$$C[(SPEC, X, Ins, Outs, \varphi)] =$$

$$= \lambda(in_1, \dots, \overset{p}{in_p}). \; fix(\; \lambda(in_1, \dots, \overset{q}{in_q}, z_1, \dots, \overset{m}{z_m}).$$

$$. (E[\varphi(out_1)], \dots, \overset{q}{E[\varphi(out_q)]}, E[\varphi(z_1)], \dots, \overset{m}{E[\varphi(z_m)]})) \downarrow 1..q$$

where $in_i \in Ins$, $out_i \in Outs$, $z_i \in X\text{-}Ins\text{-}Outs$, fix is the fixed point operator [Stoy77], $\downarrow$ is the tuple restriction operator and E is an auxiliary semantic function defined as:

$$E[\; c \; fby \; e \;] = \lambda t.(\; if \; t = 1 \; then \; E[c](1) \; else \; E[e](t-1) \;)$$
$$E[\; \sigma(e_1, \dots, e_n) \;] = \lambda t.(T_{SPEC})_\sigma (E[e_1](t), \dots, E[e_n])$$
$$E[\; x \;] = \lambda t.x(t)$$
$$E[\sigma] = \lambda t.(T_{SPEC})_\sigma$$
$$E[\; e \ll n \;] = \lambda t.E[e](\; ceil(t \,/\, N[n]) \;)$$
$$E[\; n \gg e \;] = \lambda t.E[e](\; t * N[n] \;)$$
$$E[\; (e_1 \| \dots \| e_n) \;] = \lambda t.(\; (E[e_1](t), \dots, E[e_n]) \downarrow ((t-1) \; mod \; n) + 1 \;)$$

where N is another auxiliary semantic function that relates a natural numeral with is natural number.

References

[BlEi97] C. Blumenröhr and D. Eisenbiegler. An efficient representation for formal synthesis, *Proc. International symposium on system synthesis, ISSS'97*, 1997.

[Delg87] C. Delgado. *Sematics of digital circuits, LNCS-285*, Springer Verlag, 1987.

[EhMa85] H. Ehrig and B. Mahr. Fundamentals of algebraic specification 1: equations and initial semantics, *EATCS Monographs on theoretical computer science, n° 6*, Springer-Verlag, 1985.

[FiFM91] S. Finn, M.P. Fourman and G. Musgrave. Interactive synthesis in higher order logic, *Proc. Workshop on the HOL theorem prover and its applications*, 1991.

[JoBo91] S.D. Johnson and B. Bose. DDD - A system for mechanized digital design derivation, *Proc. International Workshop on Formal Methods in VLSI Design*, 1991.

[John89] S.D. Johnson. *Manipulating logical organization with system factorizations, Hardware specification, verification and synthesis: mathematical aspects, LNCS-408,* Springer-Verlag, 1989.

[JoSh90] G. Jones and M. Sheeran. *Circuit design in Ruby,* Formal Methods for VLSI Design, ed. J. Staunstrup, North Holland, 1990.

[Kuma96] R. Kumar et al. Formal synthesis in circuit design - A classification and survey, *Proc. Formal methods in CAD, FMCAD'96,* 1996.

[MeHF96] J.M. Mendías, R. Hermida and M. Fernández. Algebraic support for transformational hardware allocation, *Proc. European Design & Test Conference, EDTC'96,* 1996.

[MeHF97] J.M. Mendías, R. Hermida and M. Fernández. Formal techniques for hardware allocation, *Proc. International conference on VLSI design, VLSI'97,* 1997.

[Pada88] P. Padawitz. Computing in Horn Clause Theories, *EATCS Monographs on theoretical computer science, nº 16,* Springer-Verlag, 1988.

[ShRa95] R. Sharp and O. Rasmussen. The T-Ruby design system, *Proc. Computer hardware description languages and their applications, CHDL'95,* 1995.

[Stoy77] J.E. Stoy. Denotational semantics: The Scott-Strachey approach to programming languaje theory, *The MIT Press Series in Computer Science, nº 1,* MIT Press, 1977.

[WaAs85] W.W. Wadge and E.A. Ashcroft. Lucid, the dataflow programming language, *APIC Studies in data processing, nº 22,* Academic Press, 1985.

6 OVERLAPPED SCHEDULING TECHNIQUES FOR HIGH-LEVEL SYNTHESIS AND MULTIPROCESSOR REALIZATIONS OF DSP ALGORITHMS

Sabih H. Gerez
Sonia M. Heemstra de Groot
Erwin R. Bonsma
Marc J.M. Heijligers

6.1 Introduction

Algorithms that contain computations that can be executed simultaneously, offer possibilities of exploiting the parallelism present by implementing them on appropriate hardware, such as a multiprocessor system or an application-specific integrated circuit (ASIC). Many digital signal processing (DSP) algorithms contain internal parallelism and are besides meant to be repeated infinitely (or a large number of times). These algorithms, therefore, not only have *intra-iteration* parallelism (between operations belonging to the same iteration) but *inter-iteration* parallelism (between operations belonging to different iterations) as well [Par91].

The distribution in time of operations in the realization of an algorithm is called a *schedule*. A schedule in which all operations of a single iteration should terminate before any of the next iteration can be started is called *nonoverlapped*. On the other hand, schedules that exploit inter-iteration parallelism are called *overlapped* because the execution of consecutive iterations overlap in time. The two types of schedules are illustrated in Figure 6.1. In the figure, i and $i + 1$ refer to the successive iteration numbers, while T_0 is the length of the iteration period (see Section 6.2). The set of

J.C. López et al. (eds.), Advanced Techniques for Embedded Systems Design and Test, 125-150.
© 1998 *Kluwer Academic Publishers.*

computations belonging to the same iteration have the same shading; individual computations are not shown.

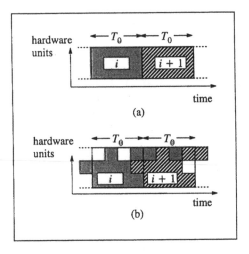

Figure 6.1. A nonoverlapped(a) and an overlapped (b) schedule

Overlapped scheduling is also called *loop folding* [Goo90] or *software pipelining* [Lam88, Jon90]. As opposed to *hardware pipelining*, that allows for overlapping the execution of subsequent operations within the same *funtional unit* (FU) by an appropriate hardware design, software pipelining overlaps the execution of a set of operations in an iteration by scheduling them appropriately on the available FUs.

This chapter deals with techniques to obtain overlapped schedules for a specific class of DSP algorithms. The chapter has a tutorial character: the main ideas are explained and references to the literature are provided. On the other hand, no experimental results are presented: they can be found in the cited literature.

The chapter is organized as follows. The first sections introduce and define the problem. Then attention is paid to the theoretical lower bound on the iteration period of an overlapped schedule. Finally, different scheduling techniques are discussed.

6.2 Data-Flow Concepts

A convenient representation for parallel computations is the *data-flow graph* (DFG). A DFG consists of *nodes* and *edges*. An edge transports a stream of *tokens* each of which carries some data value. As soon as a sufficient number of tokens is present at the input edges of a node, the node consumes these tokens and produces tokens at its output edges with values corresponding to the operation that the node is supposed to perform. Edges act as first-in first-out buffers for the transported tokens.

A DFG can represent any computation [Dav82, Eij92, Lee95]. In this chapter, however, only DFGs without conditional constructs are considered. They are sufficiently powerful to model most traditional DSP algorithms (filters). DFGs that do not contain conditional constructs are called *synchronous* [Lee87]. Synchronous DFGs have the interesting property that they can be scheduled at compile time (the number of tokens consumed and produced by each node is always known) and no run-time overhead related to data-dependent scheduling is required. If one restricts each node in a synchronous DFG only to consume a single token from each input and produce a single token on each output per invocation, one gets a *homogeneous* DFG [Lee87]. Such DFGs will be called *iterative* DFGs (IDFGs) following the terminology of [Hee92]. This chapter mainly deals with IDFGs.

An IDFG is a tuple $\langle V, E \rangle$, where the vertex set V is the set of nodes and E is the set of edges of the graph. The set V can be partitioned in a set of *computational* nodes C, a set of *delay* nodes D, a set of *input* nodes I, and a set of *output* nodes O. A computational node $c \in C$ represents an *atomic* and *nonpreemptive* computation. 'Atomic' means that each node represents a computation of the lowest level of granularity, i.e. a computation corresponding to a single activation of an FU. 'Nonpreemptive' means that a computation cannot be interrupted before completion to continue its execution at a later moment. It is further assumed that a single execution time $\delta(c)$ in the target architecture can be associated with each node c; this time is expressed in integer multiples of the system clock period. Hardware pipelined FUs are not considered in this chapter. Including them would require characterization by a second entity called the *data-initiation interval*, the time necessary to execute a pipeline stage. A delay node $d \in D$ stores a token received at its input during one *iteration period* T_0 before producing it at its output. Again, T_0 is an integer multiple of the system clock period. In a more general model, a delay node can delay a token for more than one iteration period. The number of periods that a node $d \in D$ delays its input, is indicated by $\mu(d)$. Instead of using explicit delay nodes, one could as well define delay as an edge attribute. The delay multiplicity then corresponds to the number of *initial tokens* on the edge.

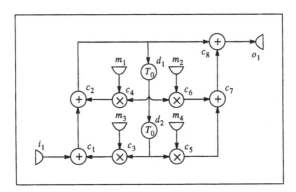

Figure 6.2. The IDFG of a second-order filter section

Although this model has some advantages from a mathematical point of view, the model with explicit delay nodes will be used here in order to be consistent with many earlier publications. An input node only produces tokens and an output node only consumes them. Figure 6.2 shows an example IDFG that represents a second-order filter section. In the figure, the node with the label T_0 represents the delay element. The nodes m_1 to m_4 that provide the constant multiplication coefficients, can be considered equivalent to inputs for the discussion in this chapter. Note that each cycle in an IDFG should at least contain a delay element. Deadlock would otherwise prevent the progress of computation in the IDFG (in terms of token flow: nodes in a cycle will never be activated if there are no initial tokens present in the cycle).

6.3. Models for the Target Architecture

Overlapped scheduling methods can be directed to different target architectures. The target architecture that is the easiest to work with from a theoretical point of view is the *ideal multiprocessor system*. It consists of a set of identical processors each of which can execute all operations in the IDFG. The time to transfer data between any pair of processors in an ideal multiprocessor system is negligible. This hardware model has been used in many studies on overlapped scheduling [Par91, Hee92, Gel93, Mad94] and will be the main model used in this chapter.

More realistic hardware models will explicitly deal with the multiprocessor network topology and the associated communication delays. Actual multiprocessor systems can roughly be divided in those that are suitable for *coarse-grain parallelism* and those suitable for *fine-grain parallelism*. In the former case, setting up a connection and transferring data will take tens of clock cycles, which means that tasks executed on the same processor should have a similar execution time in order to take advantage of the parallel hardware. Examples of scheduling approaches for DSP algorithms for this type of hardware are given in [Kon90, San96b]. Hardware that supports fine-grain parallelism, has the property that transferring a data item from one processor to the other takes just a few clock cycles. The multiprocessor system is then often integrated on a single integrated circuit [Che92, Kwe92] although systems composed of discrete commercially available digital signal processors also exist [Mad95]. A method that generates overlapped schedules for this type of architectures is discussed in Section 6.7.

Another target architecture is an *application-specific integrated circuit* (ASIC) that is the outcome of *high-level* (or *architectural*) *synthesis*. The architecture is then composed of FUs, such as adders and multipliers, for the actual calculations, registers or memories for the storage of intermediate results, and interconnection elements such as buses and multiplexers [Gaj92]. The types of architectures and clocking strategies considered normally allow the transfer of data within the same clock period as the one in which a computation terminates. This makes the scheduling problem for these architectures very similar to the one for the ideal multiprocessor system, the main difference being that FUs normally cannot execute all possible operations (additions should e.g. be mapped on adders and multiplications on multipliers). Many overlapped scheduling methods for high-level synthesis have been proposed [Goo90, Olá92, Lee94, Kos95, Wan95, Hei96].

Yet another target architecture for overlapped scheduling is a *very-long instruction word* (VLIW) processor, a processor that contains a data path with multiple FUs that can be activated in a flexible way by appropriate parts of the instruction word. It is especially in this context that the term "software pipelining" is used [Lam88].

6.4 Definitions of the Problems

In this section, first some attention is paid to the terminology related to multiprocessor scheduling and high-level synthesis. Besides, some functions are introduced that will be used in the rest of this chapter. The optimization goals to be addressed are then defined.

6.4.1 Terminology and Definitions

Multiprocessor scheduling and high-level synthesis will map each operation in the IDFG to a time instant at which the operation should start, and to an FU on which the operation will be executed (of course, many more issues, such as the storage of intermediate values in registers, should also be settled). The mapping to a time instant is called *scheduling* (in the strict sense; "scheduling" is also often used for both mappings). The scheduling will be indicated by a function $\sigma : C \rightarrow Z$ (where Z is the set of integers). For a $c \in C$, $\sigma(c)$ represents the time instance at which c starts its execution with respect to the starting time of the iteration, which is zero by definition.

The set of FUs that will be present in the final solution, is F (a processor in a multiprocessor system will also be considered to be an FU). The set of all possible operation types (e.g. addition and multiplication) is Ω. The operation type of a node is given by the function $\gamma : C \rightarrow \Omega$ and the operations that an FU can execute by the function $\Gamma : F \rightarrow 2^{\Omega}$ (2^{Ω} is the *power set* of Ω, the set of all its subsets).

The mapping of a computational node to a specific FU is called *assignment* and is given by the function $\alpha : C \rightarrow F$. Clearly, for $c \in C$, it holds that $\gamma(c) \subset \Gamma(\alpha(c))$.

Unfortunately, no uniform terminology is used in the literature. Other terms used for "assignment" include *allocation* and *binding* (while "allocation" can also mean the reservation of hardware resources for synthesis without performing yet an actual mapping). The terminology used in this chapter follows the one of [Pot92].

Note that σ and α as defined above are independent of the iteration number. This means that the type of schedules considered in this paper are *static*. Although σ maps to values in Z, time values should be taken modulo T_0 when checking for resource conflicts on the same FU. This is a direct consequence of the overlapped scheduling strategy. If the values of α depend on the iteration number in a way that the FUs to which an operation is mapped, changes according to a cyclic pattern, the schedule is called *cyclostatic* [Sch85]. Figure 6.3 shows an example of a cyclostatic schedule. In the figure, the schedules of different iterations have been given different shadings. The superscripts of the operation labels refer to the iteration number. Note that the cyclostatic pattern has a period of two. Branch-and-bound methods for obtaining cyclostatic schedules, as well as for some variants, are described in [Gel93].

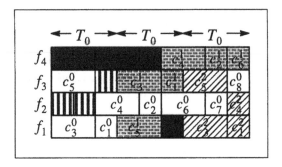

Figure 6.3. An example of a cyclostatic schedule

6.4.2 Optimization Goals

Traditionally, two optimization goals are distinguished for scheduling: one either fixes the iteration period and tries to minimize the hardware or one tries to minimize the iteration period for a given hardware configuration. The first problem is called the *time-constrained synthesis* problem, while the second is called the *resource-constrained synthesis* problem. In ASIC design, the optimization of yet a third entity, viz. *power*, is becoming more and more important. Overlapped scheduling can contribute to the reduction of power [Kim97]; this issue will, however, not be discussed further in this chapter.

In DSP, the iteration period T_0 of an algorithm is often part of the specifications and the problem to be solved is time constrained. Besides, as will become clear from the text below, T_0 should be provided in order to be able generate overlapped schedules. The resource-constrained problem can be solved by repetitively solving the time-constrained version of the problem and increasing T_0 until the resource constraints are satisfied. One can also increase T_0 during the scheduling process as is discussed in Section 6.7.

Apart from the iteration period, there is a second time entity that can be optimized. It is the *latency*, the time between the consumption of inputs and the production of the corresponding outputs. Note, that the latency can be smaller or larger than or equal to T_0 in an overlapped schedule, while the latency is never larger than T_0 in a nonoverlapped schedule.

Apart from the just mentioned "optimization versions" of the scheduling problem, also the "decision version" may be of practical importance: finding out whether a solution exists for a given set of resources (and generate a solution if one exists). This problem could be tackled in a similar way as the resource-constrained problem.

6.5 Iteration Period Bound

This section deals with the *iteration period bound* (IPB) of an IDFG, the theoretical lower bound for T_0 derived from the IDFG topology. Below, attention will be paid to the IPB in nonoverlapped and overlapped schedules, and to an efficient method to compute the IPB.

6.5.1 Nonoverlapped Scheduling

The simplest case of an IDFG is one that does not contain any delay elements and is therefore *acyclic*. When the schedule is nonoverlapped, in such a case, the IPB is given by the longest path from any of the inputs of the IDFG to any of its outputs, where path length is the sum of the execution times $\delta(c)$ of all computational nodes c in the path. This longest path is called the *critical path* of the IDFG.

Longest paths in such *directed acyclic graphs* (DAGs) can be computed in linear time, i.e. in $O(|C| + |E|)$, using well-known shortest/longest-path algorithms from computational graph theory [Cor90].

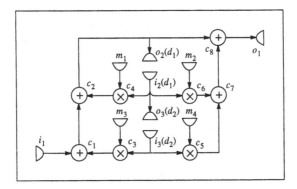

Figure 6.4. The IDFG of the second-order filter section after conversion to an acyclic graph

When the IDFG contains delay elements, these delay elements can be replaced by a pair of input and output nodes. As the scheduling is nonoverlapping anyhow, the fact that the delay element consumes a token in the current iteration and releases it in the next, can be modeled by sending the token to an output node in the current iteration and reading it from an input node in the next. Figure 6.4 shows the IDFG of Figure 6.2 after the conversion of each delay element into a pair of input and output nodes. Clearly, once that the delay nodes have been removed, the new graph is a DAG and the IPB can be computed using the linear-time longest-path algorithm mentioned above.

The fact that the IDFG contains delay elements, offers optimization possibilities using a transformation called *retiming* [Lei83, Lei91]. Retiming states that the behavior of a computation described by an IDFG does not change if delay elements connected to *all* inputs of a computational node are removed and replaced by delay elements connected to *all* outputs of the same computational node as is shown in Figure 6.5. If superscripts are again used to indicate the iteration number, one finds for Figure 6.5(a): $d^0 = a^0 + b^0$; $c^0 = d^{-1} = a^{-1} + b^{-1}$. In the case of Figure 6.5(b), one gets: $e^0 = a^{-1}$; $f^0 = b^{-1}$; $c^0 = e^0 + f^0 = a^{-1} + b^{-1}$, which shows the correctness of the transformation.

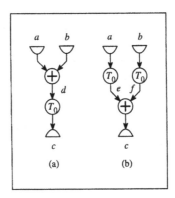

Figure 6.5. The retiming transformation

Retiming can be used to position the delay elements in the IDFG in such a way that the DAG obtained after replacing the delay elements by input-output node pairs has the smallest IPB. This can be achieved by a polynomial-time algorithm [Lei83, Lei91]. However, if resource minimization is a goal as well, the problem becomes NP-complete [Gar79] as is shown in [Pot94a].

6.5.2 Overlapped Scheduling

The value of T_0 in an IDFG without delay elements, which must be a DAG, can be arbitrarily short when overlapped schedules are allowed. New iterations can be started at any desired time, as a computation does not depend on any of the output values of earlier iterations. This is not true if the DAG was obtained by splitting each delay node into an input and output node. Then, an iteration can only start at the moment that the output values that will be used as input for the actual iteration are available.

In general, the IPB for an overlapped realization of an algorithm is given by the following expression [Rei68, Ren81]:

$$
\text{IPB} = \max_{\text{all cycles } L \text{ in the IDFG}} \frac{\sum_{c \in (L \cap C)} \delta(c)}{\sum_{d \in (L \cap D)} \mu(d)} \tag{1}
$$

This equation can be understood as follows. In a cycle of the IDFG, an operation should wait until its output has propagated back to its input before it can be executed again. This is expressed in the numerator. However, this waiting time should be distributed among the total number of delay elements present in the cycle. This is expressed by the denominator. The cycle in the IDFG for which the quotient above has the highest value determines the IPB. It is called the *critical loop*. A scheduling for which T_0 equals the IPB is called *rate optimal*. Note that the expression in *Equation 1* is independent of any retiming of the IDFG (retiming does not modify the number of delay nodes in a cycle).

In Section 6.2 it was mentioned that T_0 should be an integer multiple of the system clock period, while the result of *Equation 1* may be a fraction. In case of a fraction, one should, therefore, use the smallest integer larger than the fraction given by Equation 1. Besides, the IPB of a static schedule cannot be smaller than the largest execution time of all operations in the IDFG:

$$
\text{IPB} = \max \left(\left\lceil \max_{\text{all cycles } L \text{ in the IDFG}} \frac{\sum_{c \in (L \cap C)} \delta(c)}{\sum_{d \in (L \cap D)} \mu(d)} \right\rceil, \max_{c \in C} \delta(c) \right) \tag{2}
$$

In the rest of this chapter, the IPB as defined in Equation 2 will be used and the term "rate optimal" will refer to this definition. It is, however, possible to realize scheduling solutions that meet a fractional bound given by Equation 1 by *unfolding* [Par91] the IDFG, i.e. creating a new IDFG that represents multiple executions of the original IDFG [Jen94, San96a]. Cyclostatic scheduling (see Section 6.4.1) can also be used to achieve the IPB of *Equation 1*. Note that one can see a cyclostatic schedule as a special case of an unfolded one, viz. one with an unfolding factor equal to the cyclostatic period that should besides obey strict constraints on scheduling and assignment.

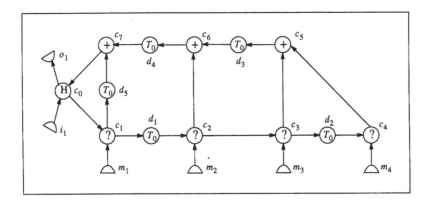

Figure 6.6. The optimally retimed correlator circuit

One might think that the IPB for a nonoverlapped schedule after optimal retiming is equal to the IPB for an overlapped schedule. This is *not* necessarily the case when the IDFG contains operations with an execution time larger than one clock period as will be illustrated by means of the example of Figure 6.6[1]. The IDFG shown is the optimally retimed correlator from [Lei83, Lei91]. The IDFG consists of computational nodes for addition, comparison (indicated by the label '?') and a host node that takes care of inputs and outputs (indicated by the label 'H'; the main function of the "dummy" host node is to impose a limit on the latency). An addition takes 7 clock periods, a comparison 3, while the host node executes in 0 clock periods. It is not difficult to see that this circuit has a critical path consisting of c_2, c_3 and c_5, which results in an IPB of 13 if the schedule should be nonoverlapped.

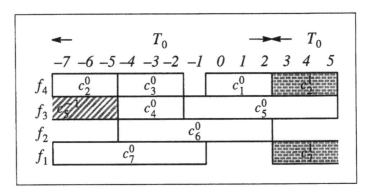

Figure 6.7. A rate-optimal overlapped schedule for the correlator

---

[1] The fact that the IPB of the optimally retimed correlator can be improved by choosing for an overlapped schedule, was discovered by András Oláh, currently employed by Ericsson, Hungary.

On the other hand, the circuit has an IPB of 10 for an overlapped schedule. This number follows from the three critical loops present (c_0 - c_1 - c_7; c_0 - c_1 - c_2 - c_6 - c_7; c_0 - c_1 - c_2 - c_3 - c_5 - c_6 - c_7). A rate-optimal schedule for an ideal multiprocessor system is shown in Figure 6.7.

It is in principle possible to realize a DSP algorithm at a speed higher than given by the IPB of *Equation 1*. This will require, however, modifications of the algorithm specifications and affect the finite word-length effects which may be quite important in DSP. Modifications involve transformations like look-ahead calculations that e.g. compute the next system state based on earlier states than the current one [Par87, Men87, Par89, Par95, Gle95].

6.5.3 Efficient Computation

The direct application of *Equation 1* (or 2) for the computation of the IPB would imply the enumeration of all cycles in the IDFG. The number of cycles can grow exponentially with respect to the number of nodes [Ger92] which implies an exponential worst-case time complexity for some graphs. Although IDFGs encountered in practice have a limited number of cycles and an enumerative approach seems to be feasible [Gel93, Wan95], it is important to notice that the problem can be solved in polynomial time. Many algorithms have been proposed for the IPB problem in IDFGs [Ger92, Kim92, Cha93, Pot94b, Ito95], some of them based on earlier solutions originally meant for other applications [Law66, Law76, Kar78]. Here, the most efficient of these methods will be explained in short.

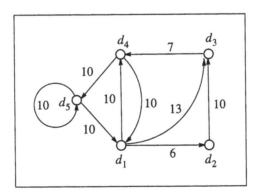

Figure 6.8. The delay graph of the optimally retimed correlator

It was originally meant to solve the *minimum cycle mean* problem [Kap78], where the "cycle mean" for a cycle in an edge-weighted graph refers to the total weight of the edges in the cycle divided by the number of edges constituting the cycle. In order to apply the minimum cycle mean method to the IPB problem for IDFGs, the original IDFG $\langle V, E \rangle$, should first be transformed into a *delay graph* $\langle D,$

E_d⟩ [Ito95], of which the vertex set consists of the delay nodes in V. The edge set E_d of this graph consists of those (u, v_j) for which there is a directed path from delay node u to delay node v that passes through computational nodes only. The edge weight $w(u, v)$ of such an edge equals the total execution time of all nodes in the path. Figure 6.8 shows the delay graph of the optimally retimed correlator of Figure 6.6. Note that the IPB is equal to the *maximal* cycle mean of the delay graph.

The maximal cycle mean algorithm, adapted from the original minimal cycle mean algorithm, consists of the following steps:

–   Choose an arbitrary node $s \in D$. Set $F_0(s) = 0$ and $F_0(v) = -\infty$, for all $v \in D$, $v \neq s$.

–   Calculate $F_k(v)$, for $k = 1, ..., |D|$ according to:

$$F_k(v) = \max_{(u,v) \in E_d} \left(F_{k-1}(u) + w(u,v) \right)$$

–   The IPB is then found from:

$$\text{IPB} = \max_{v \in D} \min_{0 \leq k \leq |D|-1} \frac{F_{|D|}(v) - F_k(v)}{|D| - k} \tag{3}$$

k	$F_k(d_1)$	$F_k(d_2)$	$F_k(d_3)$	$F_k(d_4)$	$F_k(d_5)$
0	$-\infty$	0	$-\infty$	$-\infty$	$-\infty$
1	$-\infty$	0	10	$-\infty$	$-\infty$
2	$-\infty$	0	10	17	$-\infty$
3	27	0	10	17	27
4	37	33	40	37	27
5	47	43	50	47	47

Figure 6.9. The step by step development of the maximum cycle mean algorithm

The application of this algorithm to the delay graph of Figure 6.8 is illustrated in Figure 6.9. The figure shows the values of $F_k(v)$ for all v and all k; d_2 has been chosen as the arbitrary node s with which the algorithm starts. *Equation 3* can now be applied to these values. It follows that IPB equals 10.

The construction of the delay graph from the IDFG can be achieved in $O(|D||E|)$ time using the algorithm for the construction of a *longest-path matrix* given in [Ger92]. The maximal cycle mean algorithm given above has a time complexity of $O(|D||E_d|)$. The overall time complexity of the IPB calculation method is therefore $O(|D|(|E| + |E_d|))$. Given the fact that all nodes in the IDFG have a bounded number of input edges (a multiplication or addition has e.g. two inputs), it can be stated that $|E| = O(|C|)$.

6.6 Mobility-Based Scheduling

The overlapped scheduling problem is NP-complete [Gar79, Hee92], which means that optimal solutions can only be found by algorithms that have an exponential time complexity in the worst case. Such algorithms can only be applied to small-size problems. For larger problems, one should use *heuristics* that will generate solutions that may not be optimal but can be obtained in acceptable time. One class of heuristics, the so-called *mobility-based* heuristics, are discussed in this section. First the notion of *scheduling ranges* is introduced. Then, heuristic scheduling techniques based on this notion are explained. Finally, some attention is paid to the assignment problem.

6.6.1 Scheduling Ranges

The scheduling solution as given by $\sigma(c)$ for $c \in C$ should obey in the first place the *precedence* constraints in the IDFG. For an edge $(u, v) \in E$, where $u, v \in C$, the precedence constraint implies that operation v cannot start its execution before the completion of operation u. If there is a path of n delay nodes between u and v, v cannot start before the execution of u belonging to n iterations ago is completed. In general, all precedence constraints of the IDFG for all pairs of computational nodes u and v separated by n delay nodes (n may be equal to zero) are given by:

$$\sigma(v) \geq \sigma(u) + \delta(u) - nT_0 \tag{4}$$

Note that $n = 0$ for intra-iteration precedence constraints and $n > 0$ for inter-iteration precedence constraints. All precedence constraints given by *Equation 4* can be represented in an *inequality graph* $\langle C, E_i \rangle$, using the following construction rules [Hee92]:

— The vertex set consists of the computational nodes C of the IDFG.
— Rewrite the inequalities of *Equation* 4 as:

$$\sigma(v) - \sigma(u) \geq \delta(u) - nT_0 \tag{5}$$

— The edge set E_i has a directed edge (u, v) for each pair of nodes for which an inequality as given in *Equation 4* (or 5) exists. The edge weight of the edge (u, v) is given by the right-hand side of *Equation 5*.

The inequality graph for the optimally retimed correlator of Figure 6.6 is given in Figure 6.10 when $T_0 = 10$. Note that the summation of the weights in each directed cycle is either zero (for critical loops) or negative. This is a direct consequence of the fact that T_0 should not be chosen smaller than the IPB in order for the IDFG to be computable.

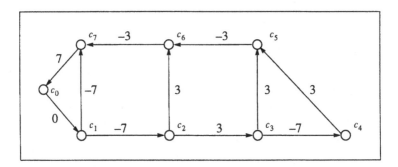

Figure 6.10. The inequality graph for the optimally retimed correlator circuit

The inequalities of *Equation 4* can only be solved if the start time in the schedule of at least one node is known. Suppose that the start time $\sigma(r)$ of some reference operation $r \in C$ is set to zero. Then, the length of the *longest path* from r to some node v in the inequality graph gives the earliest possible moment at which v is allowed to start. This time is often called the *as-soon-as-possible* (ASAP) scheduling time. On the other hand, zero minus the length of the longest path from any node v to r gives the latest possible time at which v is allowed to start. This time is called the *as-late-as-possible* (ALAP) scheduling time. For each node the ASAP and ALAP times together define its *scheduling range* [Hee92] (or *mobility interval* [Pau89]) denoted by $[\sigma_-(v), \sigma_+(v)]$. The difference $\sigma_+(v) - \sigma_-(v)$ is called the *mobility* [Pau89] or *freedom* [Par86].

The computation of the scheduling ranges amounts to the *single-source longest-path problem* and can be solved by means of the *Bellman-Ford* algorithm [Cor90] in $O(|C||E_i|)$ time (for the ALAP times one should reverse the orientation of the edges [Hee92]). Because $|E_i|$ is $O(|C|)$ (see Section 6.5.3), the complexity of computing the ranges becomes $O(|C|^2)$. The Bellman-Ford algorithm is actually an algorithm for shortest paths, but can easily be modified to compute longest paths [Sch83].

The ranges for the operations c_0 through c_7 that can be derived from the inequality graph of Figure 6.10 if c_0 is chosen as the reference node, are respectively: [0, 0], [0, 0], [-7, -7], [-4, -4], [-11, -4], [-1, -1], [-4, -4] and [-7, -7]. Note that c_4 is the only node with a mobility larger than zero because it is the only node that is not part of any critical loop. If T_0 is chosen equal to 11 instead of 10, the respective ranges become: [0, 0], [0, 1], [-8, -6], [-5, -2], [-11, -4], [-2, 1], [-5, -3], [-8, -7]. Now, all operations except for the reference have some mobility.

So far, the scheduling ranges have been determined based solely on precedence relations from the IDFG. For heuristic scheduling methods, it is important that non-optimal solutions that are allowed by these scheduling ranges, are eliminated as much as possible, without excluding any optimal solution. This increases the probability that a heuristic finds an optimal solution. Scheduling ranges can be tightened by using *schedule constraints*. A simple example is to incorporate a latency constraint. This can be accomplished by adding input and output nodes to the inequality graph, fix the inputs at time zero, and the outputs at the latency time [Kos95]. This method has also the advantage that all scheduling ranges are finite (as long as each computational node has an incoming path connecting it to an input and an outgoing path connecting it to an output, which is the case in all IDFGs that make sense).

Another possibility, reported in [Tim93a], uses resource constraints to tighten the scheduling ranges. It is based on so-called *functional unit ranges* (or *module execution intervals*), which represent the range of system clock periods in which functional units should start to execute an operation. A bipartite graph G is constructed in which edges are placed between operation scheduling ranges and functional unit ranges which have clock periods in common. A feasible schedule implies that a complete bipartite matching in G exists. Edges which can never be part of such a matching can be deleted, and the operation scheduling ranges can be tightened accordingly.

In case only a resource constraint, an iteration period constraint, or a latency constraint has been given, new constraints can be generated by using lower-bound estimation techniques [Tim93b, Rim94]. These new constraints can be used to tighten scheduling ranges even more.

6.6.2 Mobility-Based Scheduling Heuristics

Below, the principles of mobility-based scheduling methods are explained without going into the details of the methods themselves. A few methods are mentioned and some attention is paid to an efficient way of "updating" scheduling ranges.

```
"determine the scheduling ranges";
repeat
    "select an unscheduled operation c according to some rule";
    repeat
        "select a time instance t in c's scheduling range
          according to some rule";
        if "the selection satisfies all schdule constraints"
        then
            "schedule c at t";
        else
            "remove t from c's scheduling range";
            "update the scheduling ranges of the other operations";
        fi
    until "c has been scheduled"
until "all oprations have been scheduled";
```

Figure 6.11. The pseudo-code of a generic mobility-based scheduling heuristic

Given the fact that operations have mobility, one can say that the goal of the (time-constrained) scheduling problem is to fix each operation within its scheduling range in such a way that the resources necessary to implement the resulting schedule are minimized. The operations cannot be moved independently within their scheduling ranges: moving one operation may constrain the ranges of other operations. The pseudo-code of a generic mobility-based scheduling heuristic is given in Figure 6.11. Of course, the algorithm starts with the determination of the scheduling ranges applying the techniques mentioned in Section 6.6.1 and using as many constraints as possible. Many possibilities exist for selecting c and t in the algorithm. Examples of some algorithms are mentioned below. Two key activities in the algorithm, the check for scheduling constraint satisfaction and the updating of the ranges, determine the effectivity but also the complexity of the algorithm.

Mobility-based heuristics can both be used for nonoverlapped and overlapped scheduling and a heuristic developed for nonoverlapped scheduling [Par86] can often be easily adapted for the case of overlapped scheduling. A difference is that the computation of overlapped schedules requires that the resource requirements are computed after taking all times modulo T_0. Besides, the updating of scheduling ranges is more complex in the case of overlapped scheduling.

The *force-directed* scheduling method [Pau89, Ver91, Ver92], is an example of an algorithm that originally was developed for nonoverlapped schedules. It can easily be adapted for overlapped scheduling [Olá92]. The method is computationally quite intensive as it investigates many different possibilities before taking a decision on fixing or constraining a single operation.

Greedy methods, on the other hand, take decisions on fixing an operation within its range without investigating many alternatives [Hee92, Kos95]. They have the advantage of a low computational complexity, but may generate solutions of lower quality.

Recently, the combination of *genetic algorithms* [Gol89, Dav91] and greedy heuristics have shown to give good results [Hei95, Hei96]. The idea is to have a greedy heuristic that can be controlled by a permutation of the operations in the IDFG. Simply stated, whenever the heuristic should select an operation to schedule among a set of candidates, it will choose the one that is first in the permutation. It is the task of the genetic algorithm to generate different permutations. For the genetic algorithm, the greedy heuristic is just an evaluation function that returns the cost of a permutation and generates a schedule as a side effect. Some more information on this method is given in Section 6.7 when discussing a generalization of this method target for architectures with communication delays.

All mobility-based heuristic have in common that scheduling ranges should be updated after fixing an operations start time or constraining its scheduling range. This can in principle be done by recomputing all ranges using the Bellman-Ford algorithm [Hee92] as mentioned in Section 6.6.1 (of course, the algorithm should be modified to respect the scheduling decisions already taken, which is straightforward). Then, each update will have a time complexity of $O(|C|^2)$ and any mobility-based scheduling algorithm will have at least a time complexity of $O(|C|^3)$ as an update calculation is necessary after each scheduling decision.

The complexity of the range-updating problem can be reduced to $O(|C|)$ by means of a method presented in [Hei96] (similar ideas can also be found in [Lam89], where methods are used that deal with symbolic expressions in T_0). The main idea is to solve the *all-pairs longest-path* problem (see [Hee90] on how well-known shortest-path algorithms [Cor90] can be straightforwardly adapted for longest-path computations) for the inequality graph and store the results in a *inequality distance matrix* $\mathbf{D}_i[u,v]$, for all pairs $u, v \in C$. Each time that the range of some operation u is constrained by modifying either $\sigma_-(u)$ or $\sigma_+(u)$ (both are assigned the same value when the operation is fixed completely), the ranges of the remaining, yet unscheduled operations v can be updated as follows:

$$\sigma_-(v) \leftarrow \max\big(\sigma_-(v), \sigma_-(u) + \mathbf{D}_i[u,v]\big)$$

$$\sigma_+(v) \leftarrow \min\big(\sigma_+(v), \sigma_+(u) - \mathbf{D}_i[v,u]\big)$$

Clearly, a single update can be done in constant time and updating the ranges of all yet unscheduled operations can be done in $O(|C|)$ time. The initial effort to compute the inequality distance matrix can be limited to $O(|C|^2 \log|C|)$ using Johnson's algorithm [Cor90]. Actually, the matrix can be reused during different runs of a greedy algorithm, making this method of updating even more interesting than the one based on the Bellman-Ford algorithm.

Until now, it has been assumed that each operation in one iteration of the IDFG is executed exactly once. In case operations are enclosed by loop constructs, operations are executed multiple times during one iteration of the IDFG. In that case, *streams* [Mee93a] can be used to represent a particular sequence of executions of an operation, which are characterized by a vector representation. Scheduling these streams is defined as scheduling the first operation of such a stream inside its scheduling range, together with an extended constraint satisfaction and schedule range update, based upon specialized ILP techniques [Ver95, Ver97]. These methods are used in the *Phideo* high-level synthesis tool [Mee93a, Mee95].

6.6.3 Assignment

Scheduling algorithms try to minimize the number of resources to be used by the final solution e.g. keeping track of the number of operations that have to be executed simultaneously. In this section, some attention is paid on how to compute the actual number of resources required by mapping the operations on FUs (for the sake of simplicity, the ideal multiprocessor target architecture is assumed; if distinct FU types are present, the assignment problem should be solved separately for each type).

First the case of nonoverlapping scheduling is considered. After the completion of scheduling, it is known that all operations $c \in C$ occupy some FU during an *execution interval* $[\sigma(c), \sigma(c) + \delta(c) - 1]$. Operations whose intervals overlap must be assigned to distinct FUs. This problem can be modeled by a *conflict graph* $\langle C, E_c \rangle$ [Spr94] with as vertex set the set of computational nodes C and edges between those computational nodes whose execution intervals overlap. Because of the way it is constructed such a graph is called an *interval graph* [Gol80]. Suppose e.g. that the following intervals are given: $i_1 = [1, 4]$, $i_2 = [12, 15]$, $i_3 = [7, 13]$, $i_4 = [3, 8]$, $i_5 = [5, 10]$, $i_6 = [2, 6]$ and $i_7 = [9, 14]$. From these, the interval graph given in Figure 6.12(a) can be obtained.

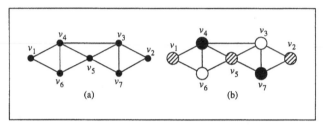

Figure 6.12. An interval graph (a) and its optimal coloring (b)

For any conflict graph, its *minimal vertex coloring* will result in optimal solution of the assignment problem. The vertex coloring problem for graphs in general is intractable [Gar79]. However, as a consequence of the way they are constructed, interval graphs form a special subset of all possible graphs that can be colored optimally in polynomial time by the *left-edge* algorithm, an algorithm that was originally published in the context of printed-circuit board routing [Has71]. The algorithm always finds a solution that uses as many FUs as the number of operations that are executed simultaneously. An optimal solution of the example just presented is shown in Figure 6.12(b).

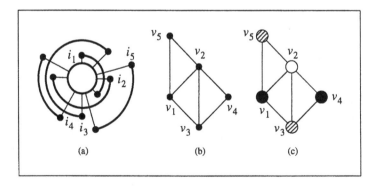

Figure 6.13. A set of circular arcs (a), its corresponding conflict graph (b) and the graph's optimal coloring (c)

In the case of overlapped scheduling, the FU occupancy cannot be modeled by linear intervals. Instead, *circular arcs* can be used to model the fact that occupancy may fold around the T_0 boundary. These arcs can be used to construct a conflict graph that is called a *circular-arc* graph. Figure 6.13 shows an example set of circular arcs, the corresponding circular-arc graph and a solution of the minimal coloring problem for this graph.

Unfortunately, the vertex coloring problem for circular-arc graphs is an NP-complete problem [Gar80]. Besides, a solution with a number of colors equal to the lower bound (the number of simultaneously executing FUs) may not exist. An algorithm that always finds the optimal solution within acceptable time for practical problems is presented in [Sto92]. A heuristic that performs very well in practice is described in [Hen93]. Both algorithms have been designed for the *register assignment* problem, the problem of mapping intermediate values that need to be stored during execution to a minimal number of registers. Note that even when nonoverlapped schedules are used for an iterative computation, the register occupancy will cross the T_0 boundary if an iteration needs results computed in previous iterations. Therefore, the register assignment problem amounts to circular-arc graph coloring even when nonoverlapped scheduling is used.

Another algorithm that also was originally developed for register assignment is described in [Mee93b]. It generates a cyclostatic assignment and in this way can deal with operations that have a longer execution time than T_0. Besides, the algorithm has the pleasant property that it guarantees a solution with at most one more FU than the lower bound.

6.7 Target Architectures with Communication Delays

This section will present the principles of an overlapped scheduling method for a target architecture with nonnegligible communication delays suitable for fine-grain parallelism. A more detailed description of the approach can be found in [Bon97] (other approaches for the same problem are unknown to the authors).

The problem is resource constrained. The resources are given by an interconnection network graph $\langle F, L \rangle$ where F is the set of FUs and L is the set of links. Each link can transfer a single data item at a time and requires a time λ for the transport.

The solution is based on a layered approach similar to the genetic scheduling method proposed in [Hei95, Hei96] and shortly discussed in Section 6.6.2. The top-level consists of a genetic algorithm that generates permutations to control a greedy lower-level heuristic. However, the greedy heuristic itself consists of two layers: a *global heuristic* that deals with an abstraction of the interconnection network and a *black-box heuristic* that refines the decisions taken by the global heuristic by routing data across the network, taking link occupancy into account, etc. The method is illustrated in Figure 6.14. The method performs scheduling and assignment simultaneously. In the presence of communication delays, the scheduling needs assignment information in order to take sensible decisions.

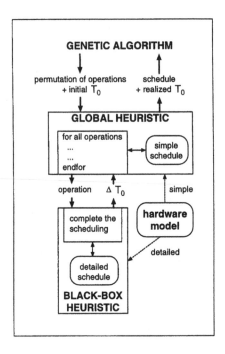

Figure 6.14. The three-layered scheduling approach

The scheduling method is based on mobility. However, the scheduling range as defined in Section 6.6.2 is not sufficient in the presence of communication delays. In order to deal with delays, the global heuristic not only uses the "inequality distance matrix" $\mathbf{D}_i$, but as well a *hardware distance matrix* $\mathbf{D}_h$. An entry $\mathbf{D}_h[f, g]$ of the matrix for two FUs $f, g \in F$ gives the shortest communication distance in system clock periods from f to g. Suppose that the computational nodes that already have been scheduled, are contained in the set S. Then the scheduling range $[\sigma_-(v), \sigma_+(v)]$ for the (tentative) assignment of a computational node v to an FU f is given by:

$$\sigma_-(v) = \max_{s \in S}\left(\sigma(s) + \mathbf{D}_i[s, v] + \mathbf{D}_h[\alpha(s), f]\right)$$

$$\sigma_+(v) = \min_{s \in S}\left(\sigma(s) - \mathbf{D}_i[v, s] - \mathbf{D}_h[f, \alpha(s)]\right)$$

It may now happen that a range is empty, i.e. that communication delays prevent the satisfaction of all precedence constraints. It may even happen that the ranges of all yet unscheduled operations for all possible assignments are empty. In such a case the algorithm performs *cycle insertion*, it increases T_0 with the goal to create valid scheduling ranges when necessary. In order not to violate the greedy character of the algorithm, the already scheduled operations are not rescheduled. This illustrated in Figure 6.15.

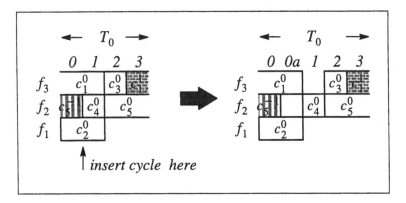

Figure 6.15. An example of cycle insertion

Both the global heuristic as well as the black-box heuristic can insert cycles when confronted with the impossibility of continuing scheduling and assignment. Clearly, it is the task of the genetic top layer to overcome the greedy nature of both heuristics and to try to find a globally optimal solution.

6.8 Conclusions

This chapter has introduced the notion of *overlapped scheduling*. As opposed to nonoverlapped scheduling, overlapped scheduling takes advantage of the parallelism present between multiple iterations in a repetitive algorithm. Many issues relevant to overlapped scheduling have been explained and a selection of important techniques has been presented.

References

[Bon97] Bonsma, E.R. and S.H. Gerez, "A Genetic Approach to the Overlapped Scheduling of Iterative Data-Flow Graphs for Target Architectures with Communication Delays", *ProRISC Workshop on Circuits, Systems and Signal Processing*, Mierlo, The Netherlands, (November 1997).

[Cha93] Chao, D.Y. and D.T. Wang, *Iteration Bounds of Single-Rate Data Flow Graphs for Concurrent Processing*, IEEE Transactions on Circuits and Systems I: Fundamental Theory and Applications, Vol. 40(9), pp. 629-634, (September 1993).

[Che92] Chen, D.C. and J.M. Rabaey, *A Reconfigurable Multiprocessor IC for Rapid Prototyping of Algorithmic-Specific High-Speed DSP Data Paths*, IEEE Journal of Solid-State Circuits, Vol. 27(12), pp. 1895-1904, (December 1992).

[Cor90] Cormen, T.H., C.E. Leiserson and R.L. Rivest, Introduction to Algorithms, MIT Press, Cambridge, Massachusetts, (1990).

[Dav82] Davis, A.L. and R.M. Keller, *Data Flow Program Graphs*, IEEE Computer, pp. 26-41, (February 1982).

[Dav91] Davis, L. (Ed.), *Handbook of Genetic Algorithms*, Van Nostrand Reinhold, New York, (1991).

[Eij92] Eijndhoven, J.T.J. van and L. Stok, "A Data Flow Graph Exchange Standard", *European Conference on Design Automation, EDAC '92*, pp. 193-199, (1992).

[Gaj92] Gajski, D.D., N.D. Dutt, A.C.H. Wu and S.Y.L. Lin, *High-Level Synthesis, Introduction to Chip and System Design*, Kluwer Academic Publishers, Boston, (1992).

[Gar79] Garey, M.R. and D.S. Johnson, *Computers and Intractability, A Guide to the Theory of NP-Completeness*, W.H. Freeman and Company, San Francisco, (1979).

[Gar80] Garey, M.R., D.S. Johnson, G.L. Miller and C.H. Papadimitriou, *The Complexity of Coloring Circular Arcs and Chords*, SIAM Journal on Algebraic and Discrete Methods, Vol. 1(2), pp. 216-227, (June 1980).

[Gel93] Gelabert, P.R. and T.P. Barnwell III, *Optimal Automatic Periodic Multiprocessor Scheduler for Fully Specified Flow Graphs*, IEEE Transactions on Signal Processing, Vol. 41(2), pp. 858-888, (February 1993).

[Ger92] Gerez, S.H., S.M. Heemstra de Groot and O.E. Herrmann, *A Polynomial-Time Algorithm for the Computation of the Iteration-Period Bound in Recursive Data-Flow Graphs*, IEEE Transactions on Circuits and Systems I: Fundamental Theory and Applications, Vol. 39(1), pp. 49-52, (January 1992).

[Gle95] Glentis, G.O. and S.H. Gerez, "Very High Speed Least Squares Adaptive Multichannel Filtering", *ProRISC/IEEE Benelux Workshop on Circuits, Systems and Signal Processing*, Mierlo, The Netherlands, pp. 123-132, (March 1995).

[Gol80] Golumbic, M.C., *Algorithmic Graph Theory and Perfect Graphs*, Academic Press, New York, (1980).

[Gol89] Goldberg, D.E., *Genetic Algorithms in Search, Optimization and Machine Learning*, Addison-Wesley, Reading, Massachusetts, (1989).

[Goo90] Goossens, G., J. Rabaey, J. Vandewalle and H. De Man, *An Efficient Microcode Compiler for Application Specific DSP Processors*, IEEE Transactions on Computer-Aided Design of Integrated Circuits and Systems, Vol. 9(9), pp. 925-937, (September 1990).

[Has71] Hashimoto, A. and J. Stevens, "Wire Routing by Optimizing Channel Assignment within Large Apertures", *8th Design Automation Workshop*, pp. 155-169, (1971).

[Hee90] Heemstra de Groot, S.M., *Scheduling Techniques for Iterative Data-Flow Graphs, An Approach Based on the Range Chart*, Ph.D. Thesis, University of Twente, Department of Electrical Engineering, (December 1990).

[Hee92] Heemstra de Groot, S.M., S.H. Gerez and O.E. Herrmann, *Range-Chart-Guided Iterative Data-Flow-Graph Scheduling*, IEEE Transactions on Circuits and Systems I: Fundamental Theory and Applications, Vol. 39(5), pp. 351-364, (May 1992).

[Hei95] Heijligers, M.J.M. and J.A.G. Jess, "High-Level Synthesis Scheduling and Allocation Using Genetic Algorithms Based on Constructive Topological Scheduling Techniques", *International Conference on Evolutionary Computation*, Perth, Australia, (1995).

[Hei96] Heijligers, M.J.M., *The Application of Genetic Algorithms to High-Level Synthesis*, Ph.D. Thesis, Eindhoven University of Technology, Department of Electrical Engineering, (October 1996).

[Hen93] Hendren, L.J., G.R. Gao, E.R. Altman and C. Mukerji, *A Register Allocation Framework Based on Hierarchical Cyclic Interval Graphs*, Journal of Programming Languages, Vol. 1(3), pp. 155-185, (1993).

[Ito95] Ito, K. and K.K. Parhi, *Determining the Minimum Iteration Period of an Algorithm*, Journal of VLSI Signal Processing, Vol. 11, pp. 229-244, (1995).

[Jen94] Jeng, L.G. and L.G. Chen, *Rate-Optimal DSP Synthesis by Pipeline and Minimum Unfolding*, IEEE Transactions on Very Large Scale Integration Systems, Vol. 2(1), pp. 81-88, (March 1994).

[Jon90] Jones, R.B. and V.H. Allan, "Software Pipelining: A Comparison and Improvement", *23rd Annual Workshop on Microprogramming and Micorarchitecture, MICRO 23*, pp. 46-56, (1990).

[Kar78] Karp, R.M., *A Characterization of the Minimum Cycle Mean in a Digraph*, Discrete Mathematics, Vol. 23, pp. 309-311, (1978).

[Kim92] Kim, J.Y. and H.S. Lee, *Lower Bound of Sample Word Length in Bit/Digit Serial Architectures*, Electronics Letters, Vol. 28(1), pp. 60-62, (January 1992).

[Kim97] Kim, D. and K. Choi, "Power-Conscious High-Level Synthesis Using Loop Folding", *34th Design Automation Conference*, pp. 441-445, (1997).

[Kon90] Konstantinides, K., R.T. Kaneshiro and J.R. Tani, *Task Allocation and Scheduling Models for Multiprocessor Digital Signal Processing*, IEEE Transactions on Acoustics, Speech and Signal Processing, Vol. 38(12), pp. 2151-2161, (December 1990).

[Kos95] Koster, M.S. and S.H. Gerez, "List Scheduling for Iterative Data-Flow Graphs", *GRONICS '95, Groningen Information Technology Conference for Students*, pp. 123-130, (February 1995).

[Kwe92] Kwentus, A.Y., M.J. Werter and A.N. Willson, *A Programmable Digital Filter IC Employing Mulitple Prcessors on a Single Chip*, IEEE Transactions on Circuits and Systems for Video Technology, Vol. 2(2), pp. 231-244, (June 1992).

[Lam88] Lam, M., "Software Pipelining: An Effective Scheduling Technique for VLIW Machines", *SIGLAN '88 Conference on Programming Language Design and Implementation*, pp. 318-328, (June 1988).

[Lam89] Lam, M.S., *A Systolic Array Optimizing Compiler*, Kluwer Academic Publishers, Boston, (1989).

[Law66] Lawler, E.L., "Optimal Cycles in Doubly Weighted Directed Linear Graphs", *International Symposium on the Theory of Graphs*, Rome, pp. 209-213, (1966).

[Law76] Lawler, E.L., *Combinatorial Optimization: Networks and Matroids*, Holt, Rinehart and Winston, New York, (1976).

[Lee87] Lee, E.A. and D.G. Messerschmitt, *Synchronous Data Flow*, Proceedings of the IEEE, Vol. 75(9), pp. 1235-1245, (September 1987).

[Lee94] Lee, T.F., A.C.H. Wu, Y.L. Lin and D.D. Gajski, *A Transformation-Based Method for Loop Folding*, IEEE Transactions on Computer-Aided Design of Integrated Circuits and Systems, Vol. 13(4), pp. 439-450, (April 1994).

[Lee95] Lee, E.A. and T.M. Parks, *Dataflow Process Networks*, Proceedings of the IEEE, Vol. 83(5), pp. 773-799, (May 1995).

[Lei83] Leiserson, C.E., F.M. Rose and J.B. Saxe, "Optimizing Synchronous Circuitry by Retiming (Preliminary Version)", In: R. Bryant (Ed.), *Third Caltech Conference on VLSI*, Springer Verlag, Berlin, pp. 87-116, (1983).

[Lei91] Leiserson, C.E. and J.B. Saxe, *Retiming Sychronous Circuitry*, Algorithmica, Vol. 6, pp. 5-35, (1991).

[Mad94] Madisetti, V.K. and B.A. Curtis, *A Quantitative Methodology for Rapid Prototyping and High-Level Synthesis of Signal Processing Algorithms*, IEEE Transactions on Signal Processing, Vol. 42(11), pp. 3188-3208, (November 1994).

[Mad95] Madisetti, V.K., *VLSI Digital Signal Processors, An Introduction to Rapid Prototyping and Design Synthesis*, IEEE Press and Butterworth Heinemann, Boston, (1995).

[Mee93a] Meerbergen, J. van, P. Lippens, B. McSweeney, W. Verhaegh, A. van der Werf and A. van Zanten, *Architectural Strategies for High-Throughput Applications*, Journal of VLSI Signal Processing, Vol. 5, pp. 201-220, (1993).

[Mee93b] Meerbergen, J.L. van, P.E.R. Lippens, W.F.J. Verhaegh and A. van der Werf, "Relative Location Assignment for Repetitive Schedules", *European Conference on Design Automation with the European Event on ASIC Design, EDAC/EUROASIC*, pp. 403-407, (1993).

[Mee95] Meerbergen, J.L. van, P.E.R. Lippens, W.F.J. Verhaegh and A. van der Werf, *PHIDEO: High-Level Synthesis for High Throughput Applications*, Journal of VLSI Signal Processing, Vol. 9, pp. 89-104, (1995).

[Men87] Meng, T.H.Y. and D.G. Messerschmitt, *Arbitrarily High Sampling Rate Adaptive Filters*, IEEE Transactions on Acoustics, Speech and Signal Processing, Vol. ASSP-35(4), pp. 455-470, (April 1987).

[Olá92] Oláh, A., S.H. Gerez and S.M. Heemstra de Groot, "Scheduling and Allocation for the High-Level Synthesis of DSP Algorithms by Exploitation of Data Transfer Mobility", *International Conference on Computer Systems and Software Engineering, CompEuro 92*, pp. 145-150, (May 1992).

[Par86] Parker, A.C., J.T. Pizarro and M. Mlinar, "MAHA: A Program for Datapath Synthesis", *23rd Design Automation Conference*, pp. 461-466, (1986).

[Par87] Parhi, K.K. and D.G. Messerschmitt, *Concurrent Cellular VLSI Adaptive Filter Structures*, IEEE Transactions on Circuits and Systems, Vol. CAS-34(10), pp. 1141-1151, (October 1987).

[Par89] Parhi, K.K., *Algorithm Transformation Techniques for Concurrent Processors*, Proceedings of the IEEE, Vol. 77(12), pp. 1879-1895, (December 1989).

[Par91] Parhi, K.K. and D.G. Messerschmitt, *Static Rate-Optimal Scheduling of Iterative Data-Flow Programs via Optimum Unfolding*, IEEE Transactions on Computers, Vol. 40(2), pp. 178-195, (February 1991).

[Par95] Parhi, K.K., *High-Level Algorithm and Architecture Transformations for DSP Synthesis*, Journal of VLSI Signal Processing, Vol. 9, pp. 121-143, (1995).

[Pau89] Paulin, P.G. and J.P. Knight, *Force-Directed Scheduling for the Behavioral Synthesis of ASICs*, IEEE Transactions on Computer-Aided Design of Integrated Circuits and Systems, Vol. 8(6), pp. 661-679, (June 1989).

[Pot92] Potkonjak, M. and J.M. Rabaey, *Scheduling Algorithms for Hierarchical Data Control Flow Graphs*, International Journal of Circuit Theory and Applications, Vol. 20, pp. 217-233, (1992).

[Pot94a] Potkonjak, M. and J. Rabaey, *Optimizing Resource Utilization Using Transformations*, IEEE Transactions on Computer-Aided Design of Integrated Circuits and Systems, Vol. 13(3), pp. 277-292, (March 1994).

[Pot94b] Potkonjak, M. and J. Rabaey, *Optimizing Throughput and Resource Utilization Using Pipelining: Transformation Based Approach*, Journal of VLSI Signal Processing, Vol. 8, pp. 117-130, (1994).

[Rei68] Reiter, R., *Scheduling Parallel Computations*, Journal of the ACM, Vol. 15(4), pp. 590-599, (October 1968).

[Ren81] Renfors, M. and Y. Neuvo, *The Maximum Sampling Rate of Digital Filters Under Hardware Speed Constraints*, IEEE Transactions on Circuits and Systems, Vol. CAS-28(3), pp. 196-202, (March 1981).

[Rim94] Rim, M. and R. Jain, *Lower-Bound Performance Estimation for the High-Level Synthesis Scheduling Problem*, IEEE Transactions on Computer-Aided Design of Integrated Circuits and Systems, Vol. 13(4), pp. 451-457, (April 1994).

[San96a] Sanchez, F. and J. Cortadella, "Maximum-Throughput Software Pipelining", *2nd International Conference on Massively Parallel Computing Systems*, pp. 483-490, (May 1996).

[San96b] Sanchez, J. and H. Barral, *Multiprocessor Implementation Models for Adaptive Algorithms*, IEEE Transactions on Signal Processing, Vol. 44(9), pp. 2319-2331, (September 1996).

[Sch83] Schiele, W.L., "On a Longest Path Algorithm and Its Complexity If Applied to the Layout Compaction Problem", *European Conference on Circuit Theory and Design*, pp. 263-265, (1983).

[Sch85] Schwartz, D.A., *Synchronous Multiprocessor Realizations of Shift-Invariant Flow Graphs*, Report no. DSPL-85-2, Ph.D. Thesis, Georgia Institute of Technology, School of Electrical Engineering, (June 1985).

[Spr94] Springer, D.L. and D.E. Thomas, *Exploiting the Special Structure of Conflict and Compatibility Graphs in High-Level Synthesis*, IEEE Transactions on Computer-Aided Design of Integrated Circuits and Systems, Vol. 13(7), pp. 843-856, (July 1994).

[Sto92] Stok, L. and J.A.G. Jess, *Foreground Memory Management in Data Path Synthesis*, International Journal of Circuit Theory and Applications, Vol. 20, pp. 235-255, (1992).

[Tim93a] Timmer, A.H. and J.A.G. Jess, "Execution Interval Analysis under Resource Constraints", *International Conference on Computer-Aided Design*, pp. 454-459, (1993).

[Tim93b] Timmer, A.H., M.J.M. Heijligers, L. Stok and J.A.G. Jess, "Module Selection and Scheduling Using Unrestricted Libraries", *European Design Automation Conference (EDAC/EUROASIC '93)*, pp. 547-551, (1993).

[Ver91] Verhaegh, W.F.J., E.H.L. Aarts, J.H.M. Korst and P.E.R. Lippens, "Improved Force-Directed Scheduling", *European Design Automation Conference*, pp. 430-435, (1991).

[Ver92] Verhaegh, W.F.J., P.E.R. Lippens, E.H.L. Aarts, J.H.M. Korst, A. van der Werf and J.L. van Meerbergen, "Efficiency Improvements for Force-Directed Scheduling", *International Conference on Computer-Aided Design*, pp. 286-291, (1992).

[Ver95] Verhaegh, W.F.J., *Multidimensional Periodic Scheduling*, Ph.D. Thesis, Eindhoven University of Technology, Department of Applied Mathematics, (December 1995).

[Ver97] Verhaegh, W.F.J., P.E.R. Lippens, E.H.L. Aarts and J.L. van Meerbergen, "Multi-dimensional Periodic Scheduling: A Solution Approach", *European Design and Test Conference, ED\&TC '97*, pp. 468-474, (1997).

[Wan95] Wang, C.Y. and K.K. Parhi, *High-Level DSP Synthesis Using Concurrent Transformations, Scheduling and Allocation*, IEEE Transactions on Computer-Aided Design of Integrated Circuits and Systems, Vol. 14(3), pp. 274-295, (March 1995).

7 SYNTHESIS OF RECONFIGURABLE CONTROL DEVICES BASED ON OBJECT-ORIENTED SPECIFICATIONS

Valery Sklyarov
António Adrego da Rocha
António de Brito Ferrari

7.1 Introduction

There are many kinds of devices that can be decomposed into a **datapath** (execution unit) and **control units**. The datapath contains storage and functional units. A control unit performs a set of instructions by generating the appropriate sequence of microinstructions that depends on intermediate logic conditions or intermediate states of the datapath. Each instruction describes what operations must be applied to which operands stored in the datapath (or in external memory).

A particular kind of execution unit that can appear in embedded (reactive) control systems [Migu97, ELLV97] is a device controlled by a sequence of instructions such as switch on/off, set in given state, reset, etc. These instructions affect the components of the execution unit via actuators, and depend on the intermediate states of its sensors, storage, and functional elements. The components and sensors could be electronic, optical, mechanical, etc. Embedded systems are parts of larger systems [Migu97] and they are widely used in the manufacturing industry, in consumer products, in vehicles, in communication systems, in industrial automation, etc. [Migu97, ELLV97]. Since typical embedded systems are heterogeneous and their specifications may change continuously [ELLV97] we have to provide them with such facilities as flexibility and extensibility.

In most digital systems, the hardware functions desired could be specified and modified by some software programs, which are applied at different levels. Most

J.C. López et al. (eds.), Advanced Techniques for Embedded Systems Design and Test, 151-177.
© 1998 *Kluwer Academic Publishers.*

digital systems are based on the use of an instruction set architecture and its variety called application-specific instruction set architecture [Migu97]. Another approach is called hardware-level programming. This is used in order to configure and control the hardware in accordance with the given specification. There are many different techniques, which can be applied for such purposes, for example microprogramming. An attractive approach is based on the use of field programmable logic devices (PLD) i.e. integrated circuits that are customized in the field [Maxf96]. This makes it possible to change the structure and to configure the interconnection of hardware circuits after they have been manufactured.

The generic PLD can be divided into several groups [Maxf96], such as simple PLDs (PROM, PLA, PAL, etc.), complex PLDs (for instance, XPLA [Pirp97], which combines PLA with PAL) and field programmable gate arrays (FPGAs). It should be mentioned that the new dynamically reconfigurable FPGA, such as Xilinx XC6200 [Xili97, LySt96] might be customized on-the-fly to implement a specific software function with very high performance. They can be reprogrammed partially without suspending operation of other parts that do not need to be reconfigured [Xili97]. This is achieved by specifying the desired interconnections between logic components (gates, flip-flops, etc.) using SRAM. Dynamic partial reconfiguration is implemented by defining a "window" in the memory map at a specific X,Y location, which can be loaded and updated on-the-fly [Maye97].

The designer of digital systems (especially embedded systems) must pay attention to such problems as the addition of new functionality in the future, and satisfaction of real time constraint [ELLV97]. Sometimes, in order to shorten the development lead-time (or possibly for other reasons), the designer will want to explore alternative realizations of some components that can be implemented in the first release of the product in order to start production, and that can later be re-evaluated and improved in order to increase performance, or to decrease cost [Migu97]. It is also very important to provide for the reuse of hardware and software macro blocks, which reduces design and development time as compared to traditional circuit design methodologies, and can lead to products of superior quality [Migu97].

Flexibility, extensibility and reuse can be achieved in digital systems based on reprogrammable and reconfigurable components that exploit field PLD technology, and can be personalized after manufacturing to fit a specific application.

Most models of computation in digital systems and in embedded systems in particular include components with state, where behavior is given as a sequence of state transitions [ELLV97], which is served by control circuits. In order to provide a system with the properties we considered above, we have to be able to reprogram and reconfigure control circuits after manufacturing.

The chapter is organized in six sections. Section 7.1 is this introduction. Section 7.2 presents a brief overview of different techniques, which might be applied to behavioral specification and logic synthesis of control units. Section 7.3 shows how to describe the behavior of control devices by graph-schemes and their varieties.

Section 7.4 discusses various methods of logic synthesis, which can be applied to the design of control circuits whose behavior have been specified by hierarchical graph-schemes. Section 7.5 presents different architectures of control circuits that are based on reprogrammable elements and reconfigurable structures. It also explains how the methods in section 7.4 can be adapted. Section 7.6 suggests combining of methods of digital design with the technique of object-oriented programming.

7.2 An Overview of Different Approaches to Behavioral Description and Logic Synthesis of Control Devices

Consider the following problem: for a given set of instructions $\pi=\{\pi_1,...,\pi_k\}$ and constraints $\phi=\{\phi_1,...,\phi_p\}$, design a control unit which will perform π and satisfy the set of conditions ϕ. The set ϕ specifies logic primitives (building blocks) with their physical parameters (such as the number of inputs, the number of outputs, etc.), structural constraints (such as the maximum number of logic primitives available for a particular circuit), timing constraints, etc.

There are many methods of logic synthesis that can be applied to solve this general problem [Bara94, BaSk86, Skl84a], etc. Suppose it is also necessary to broaden our problem. We want to construct a reconfigurable and reprogrammable scheme that contains elements with changeable functions that are initially undefined. We will call this scheme a basic scheme. Most of the connections between elements are generally fixed and they cannot be changed. Basically each particular scheme can be considered as a template for an infinite number of different applications. The customizing of the base scheme (implementing a particular control algorithm) is carried out by programming (or reprogramming) its elements with changeable functions. We are also assuming that a small number of the connections between elements can be changed, either to satisfy the requirements defined by a given set ϕ, or to improve some of the application-specific parameters.

Using base schemes allows us to provide a structured approach to the control unit design. These schemes can be implemented in FPGAs. As a result, we will be able to optimize a chip for a specific class of circuits, such as control devices.

Similar approaches were considered in [ChLe96, WRVr95]. These approaches make use of FPGA architectures that take advantage of specific circuits [Brow96], such as datapaths [ChLe96] and memory [WRVr95]. An overview of different circuit-specific methods of FPGA design was presented in [Brow96]. The complexity of recent FPGAs allows for the implementation of complete systems, including memory, datapath, interface and control units, so having areas of an FPGA dedicated to certain types of circuits would make sense [Bost96].

The main purpose of the approach we are considering is to provide extensibility, flexibility and reuse of the schemes for different control units. This implies accommodating the following additional requirements:

- allow the set of instructions π to be extended;

- allow the instructions in the set π to be changed;

- enable previously designed components of the control unit and previously designed fragments of the control algorithms to be used for future applications without redesigning them.

It should be mentioned that any element π_i of π, can be seen as a description of the component with states whose behavior has been specified as a sequence of predefined state transitions that depend on predefined conditions. In this sense, the approach can be applied to the formal synthesis of a wide range of global and local control circuits.

Most of the relevant work on this is in the area of virtual application-specific circuits that offer a mixture of hardware performance and software versatility and onward as application-specific co-processors in general-purpose computing systems. This is the approach that has been used in PRISM developed at Brown University [AtSi93], Splash, developed at the Supercomputer Research Centre in USA [Gokh91], Spyder, developed at EPFL in Switzerland [IsSa93] and PAM [VBRS96], developed at the Digital Equipment Research Laboratories in Paris. The proposal is for a standard high-performance microprocessor enhanced by a PAM (programmable active memories) coprocessor, which consists of a board of FPGA's and local memory interfaces to a host computer. The purpose of PAM is to implement a virtual machine that can be dynamically reconfigured as a broad class of specific hardware devices, one for each interesting application. For example, the performance-critical portion of the program can be extracted and implemented in an FPGA-based board, which essentially increases the speed of execution. Generally speaking, PAM's merge hardware and software together, and they are based on virtual circuits such as FPGAs. The paper [VBRS96] also presents a good overview of reconfigurable devices, their structures and components.

The popular approach is a design with cores that reduces the complexity of the design process and decreases the development time. The paper [ZhPa96] presents a brief overview of research in this area and a method for generating a reuse-friendly RTL structure based on multiple behaviors.

An attractive approach to reconfigurable circuit design borrows some ideas from the object-oriented model [ChLR93, Kuma94, Hura97]. The authors of the paper [ChLR93] investigate the benefits of using such ideas for modeling digital scheme design information. The design types considered in [ChLR93] have two inter-related views - the interface and the implementation. These can be configured statically (where there is an explicit association between an implementation and the interface) and dynamically (where an implementation is selected according to some criterion). As a result, some important concepts from object-oriented programming such as data abstraction and inheritance can be invoked. The authors of the paper [Kuma94] identify reusable hardware components using a data decomposition approach. They illustrate the benefits of object-oriented techniques in such areas as reuse, and the specialization of components through inheritance, as well as the identification of reusable components through data decomposition. The paper [Hura97] shows how object-oriented methodology can be applied to medium and large real time distributed embedded systems in order to provide flexibility, extensibility, reusability and maintainability of design and implementation. The approach is based on the object model and the message passing mechanism. The paper [Skly96] is devoted to the design of reusable hardware components that can be adapted to construct finite state machines (FSM) for different application-specific tasks.

All these and other related papers have focused on particular design tasks that are included in the general area of designing extensible, flexible and reusable digital circuits. This chapter is devoted to the synthesis of control units that have these properties. It differs from other related work, not only in the specific application area, but also in the approach. The technique used enables us to provide a formal and complete synthesis from a given behavioral description for an application-specific virtual control device which can be reprogrammed for either new or modified versions of the original behavioral description.

There are many different methods and tools, which can be used in order to describe a behavior of control units. Since digital control can be expressed as a sequence of states and state transitions we can apply any behavioral description of FSM, such as state transition graphs, state transition tables, etc. A higher level of abstraction can be achieved in the use of operational schemes of algorithms, such as graph-schemes, logic schemes and matrix schemes [Bara94]. These tools enable us to separate the interface of the control unit from its internal implementation. They have also some other advantages, considered in [Bara94].

Graph-schemes (GS) (or binary decision trees) and their varieties (extended graph-schemes, hierarchical graph-schemes, parallel hierarchical graph-schemes) are used below for the behavioral description of control units.

Many different methods of logic synthesis of control circuits [Bara94, BaSk86, Skl84a, Mich94, Katz94], etc. have been developed. Mainly they are based on the use of FSM theory and the microprogramming approach. FSM is the general mathematical model of a control unit. The technique involves the traditional steps, such as state minimization, state encoding, Boolean function optimization, etc. In case of the microprogramming approach various encoding strategies are applied.

The technique used in the chapter on the one hand mainly follows traditional steps but on the other hand takes into consideration different criteria of optimization, which finally allow us to design reprogrammable and reconfigurable virtual circuit directly supporting top-down decomposition of the control algorithm.

7.3 Graph-Schemes and their Varieties

Graph-Schemes (GS) of algorithms (see an example in figure 7.1) are used in order to describe the behavior of control circuits [Bara94]. They have the following formal description:

- a GS is a directed connected graph, which is composed of rectangular and rhomboidal nodes. Each GS has one entry point which is a rectangular node marked with a *Begin* label, and one exit point which is a rectangular node marked with an *End* label;

- other rectangular nodes contain microinstructions from the set $\iota=\{Y_0,Y_1,Y_2,...\}$. Any microinstruction Y_j, includes a subset of microoperations from the set $Y=\{y_1,...,y_N\}$. A microoperation is an output signal, which causes a simple action in the datapath such as setting a register or incrementing a counter;

- each rhomboidal node contains just one element from the set X, where $X=\{x_1,...,x_L\}$ is the set of logic conditions. A logic condition is an input signal, which communicates the result of a test, such as the state of a sensor;

- all nodes, except the node *Begin*, have only one input. The node *Begin* has no inputs. All rectangular nodes, apart from the node *End*, have only one output, and a rhomboidal node has two outputs marked with "1" (true) and "0" (false). The *End* node has no outputs;

- inputs and outputs of the nodes are connected by directed lines (arcs), which go from the output to the input in such a way that:

- every output is connected with only one input;

- every input is connected with at least one output;

- every node is located on at least one of the paths, which go from the node *Begin* to the node *End*.

Various kinds of GSs allow us to describe sequential and parallel devices, the duration of clock pulses for synchronization, the hierarchical ordering of operations to be performed, etc. They provide good separation of the control unit's interface from its implementation. Figure 7.1 shows an example of a graph-scheme, which describes the behavior of a control circuit with 4 inputs $(x_1,...,x_4)$ and 6 outputs $(y_1,...,y_6)$.

Extended GSs allow specifying for each microinstruction the particular time of its execution. For example, different timing intervals have been designated in figure 7.1 by values of the variables $t_1,...,t_3$ and these values are different. The method of synthesis of control circuits, described by extended GS, was considered in [KiSk79].

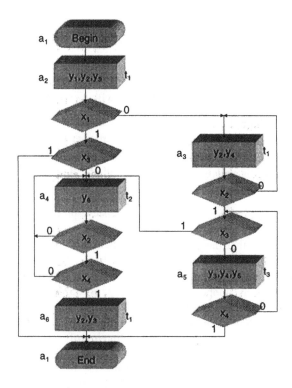

Figure 7.1. An example of a graph-scheme

Hierarchical GS (HGS) allows to specify hierarchical ordering of operations to be performed. As a result a hierarchical algorithm can be viewed at various levels of abstraction. HGSs are GSs with the following distinctive features [Skl84a, Skl84b]:

- their rectangular nodes contain either microinstructions from the set $\iota=\{Y_0,Y_1,Y_2,...\}$ or macroinstructions from the set $\varphi=\{\}_1,\}_2,...\}$, or both. Any macroinstruction incorporates a subset of macrooperations from the set $Z=\{z_1,...,z_Q\}$. Each macrooperation is described by another HGS of a lower level. For the beginning we assume that each macroinstruction includes just one macrooperation, meaning that we are considering only sequential (non-parallel) processes. Later we will eliminate this restriction

- their rhomboidal nodes contain just one element from the set $X\cup\Theta$, where $X=\{x_1,...,x_L\}$ is the set of logic conditions, and $\Theta=\{\theta_1,...,\theta_I\}$ is the set of logic functions. Each logic function is calculated by performing some predefined set of sequential steps that are described by an HGS of a lower level.

Consider the set $E=\{\varepsilon_1,...,\varepsilon_V\}$, for which $E=Z\cup\Theta$. Each element $\varepsilon_v\in E$ corresponds to HGS Γ_v, which describes either an algorithm for performing ε_v (if $\varepsilon_v\in Z$) or an algorithm for calculating ε_v (if $\varepsilon_v\in\Theta$). In both cases an algorithm is being described using HGS of lower level. Let us assume that $Z(\Gamma_v)$ is the subset of macrooperations and $\Theta(\Gamma_v)$ is the subset of logic functions that belong to HGS Γ_v. If $Z(\Gamma_v) \cup \Theta(\Gamma_v) = \emptyset$ we have an ordinary GS [Bara94] or a one-level representation of an algorithm.

In order to prevent infinite recursion in execution of HGSs they must be checked for correctness using the technique [Skl84a]. We will consider only correct HGSs.

A **hierarchical algorithm** of a logic control can be expressed by a set of HGSs, which describe the main part and all elements of the set E. The main part is being described by HGS Γ_1 from which the execution of the control algorithm will be started. All other HGSs will be subsequently called either from Γ_1 or from other HGSs that are descendants of Γ_1.

Any HGS has state, behavior and identity. The same properties have been used in order to characterize software objects in object-oriented programming (OOP) [Booc94]. It is known that there is an analogy between software objects and finite state machines (FSM) [Booc94]. OOP has been defined as a method of implementation in which programs are organized as cooperative collections of objects [Booc94] or as cooperative collections of FSMs. The actual boundary between advantages of FSM specification and object-oriented specification is fuzzy. For some objects in OOP the event and time ordering of operations is so important that we might best formally describe the behavior of such objects using FSM notation [Booc94]. On the other hand for many hardware applications we can apply object-oriented notation. Since HGSs enable us to specify a behavior of communicating FSM or a set of communicating hardware objects (see section 7.6) we can visualize this notation as a kind object-oriented specification. Indeed it enables us to describe a behavior of individual FSM (that is object) and a behavior of cooperative collections of FSMs (that are objects). In section 7.6 we will show how to apply this notation to describe a behavior of hardware objects.

Using HGSs enables us to develop any complex control algorithm part by part concentrating our efforts on different levels of abstraction. Each component of the E can be independently tested and debugged. It is well known that there is only one basic way of dealing with complexity: "Divide and conquer". This well-known idea can be applied to a variety of objects, for instance, to HGSs that simplify the design of complex control algorithms. Figure 7.2 demonstrates a description of an algorithm by HGSs Γ_1, Γ_2, Γ_3, Γ_4 and $Z = \{Z_1, Z_2, Z_3, Z_4\}$, $\Theta = \emptyset$.

Some operations in an HGS can be designated as virtual. A macrooperation (a logic function) is called **virtual** if it is not permanently attached during the design of a control unit. For any virtual element (VE), which is either a macrooperation or a logic function, we can accept in future different implementations. VE is described by the appropriate virtual HGS. The virtual HGS can be seen as a variable part of the control algorithm.

A virtual element is called a **pure virtual** element (PVE) if it just has a name and does not have any implementation. PVE is described by a virtual HGS, which is composed of just two nodes following each other: **Begin** and **End**. The notions considered above were borrowed from OOP [Booc94].

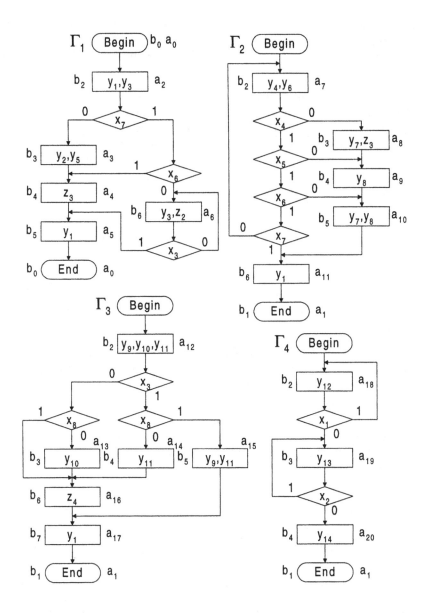

Figure 7.2. A description of an algorithm by hierarchical graph-schemes

If a HGS has at least one PVE, it is called an incomplete HGS. An incomplete HGS can be executed for the purposes of testing, but ultimately all PVEs have to be replaced with non pure VEs. VE is indicated by a superscript v, for example $z_q{}^v$, $\theta_i{}^v$. PVE is indicated by a superscript pv, for example $z_q{}^{pv}$, $\theta_i{}^{pv}$.

Parallel HGS (PHGS) includes macroinstructions which are composed of more than one macrooperation. When a macroinstruction has more than one macrooperation all of them will be executed in parallel [Skly87]. The transition from any active rectangular node to the next node will be carried out when all its components (a microinstruction and/or a macroinstruction) will be terminated.

The execution of an ordinary graph-scheme based on a special traversal procedure was considered in [Bara94]. We will use a similar approach that differs only in the interpretation of new complex operations such as $\varepsilon_v = z_q \in Z$ and $\varepsilon_v = \theta_i \in \Theta$ ($v \in \{1,...,V\}$). Each complex operation ε_v, described by a separate HGS Γ_v, initiates the execution of a new sub-sequence of operators. When a macroinstruction contains more than one macrooperation, the interrupted HGS will be resumed after all parallel macrooperations will have been executed.

7.4 Synthesis of Control Units Described by Hierarchical Graph-Schemes

By synthesis we mean a process of design refinement where a more abstract specification, such as a GS, is translated into a less abstract specification, such as a logic scheme built from given elements (logic primitives).

If a control algorithm is described by a GS [Bara94] we can apply well known methods of logic synthesis [Bara94, Skl84a, Skly96]. They consider two targets of the design, which are the following:

- the scheme built from reprogrammable elements (RE). A particular RE is used to implement a system of Boolean functions with predefined constraints, such as the maximum number of arguments, the maximum number of functions in the system, etc. Each component could be either a part of a PLD or a part of a more complex structure, such as an FPGA;

- the scheme built from gates. This task has to be solved when our control circuit will be later implemented in a gate-based structure, such as FPGAs with fine-grain architecture [Xili97, LySt96, Maye97].

For instance the approach proposed in [BaSk86, Skl84a, Skly96] is based on the following steps of logic synthesis:

- Converting a GS to structural tables that are used in a similar way to state transition diagrams of a FSM. The objective is to satisfy the given constraints, such as the actual numbers of inputs and outputs of REs. Here RE is either a reprogrammable element for the first target considered above or a reconfigurable element for the second target. For a given GS Γ (see, for instance, figure 7.1) and RE(n,m) (where n and m are the number of inputs and the number of outputs of the RE respectively) it is necessary to build a set of structural sub-tables such that: they describe the total behavior represented by Γ; the number of sub-tables is minimal; each sub-table contains information for programming (configuring) a corresponding RE (it means that all constraints have been satisfied). The methods [BaSk86, Skl84a, Skly96] are based on marking the given GS with labels, which will be used to name states [Bara94] and partitioning of state transitions in accordance with predefined relations. If the problem persists the technique of state splitting [BaSk86, Skl84a] is applied.

- State encoding. The objective is to reduce the functional dependency of outputs on inputs. This will allow us to decrease the number of output lines used for transition functions $D_1,...,D_R$ in each RE, and to simplify the scheme of the control circuit.

- Combinational logic optimization and designing the final scheme. The main ideas are based on a special decomposition aimed at utilizing predefined reusable frames in spite of constructing schemes with an arbitrary structure.

The distinctive feature of this and many other methods is that FSM states are considered to be an internal attribute of the scheme. If we need to redesign the scheme we have to modify many parameters and essentially change its logic structure. Since state encoding radically affects the structure of the logic components and their interconnections, it is difficult to provide for modifications after the scheme has been synthesized.

An alternative approach to the design of control circuits is based on the model, in which the output register of the FSM is simultaneously used as the state register. As a result the size of the register will be larger than that required when ordinary state encoding techniques are used. In practice this is not a problem because the cost of a flip-flop is negligible in FPGA based circuits. For example, each cell in the XC6200 FPGA family can be configured as a two-input logic primitive, or as a D flip-flop, or as a primitive connected to the D flip-flop [Xili97, LySt96, Maye97]. So we can combine a sample combinational logic with elements of the register (with D flip-flops) inside the same cell. A problem can arise when we want to generate the same microinstructions in different states. However adding the minimal number of new flip-flops in order to recognize the proper state can trivially solve it.

Independently on model of the control circuit to be used, the GS can be directly transformed to the logic scheme. Since branches of any GS present binary decisions for feasible transitions between rectangular nodes we can provide an association between these branches and gate-based sub-schemes, which logically calculate the next active rectangular node (i.e. the next microinstruction). In order to select active path in GS we have to calculate Boolean equations, which are mainly composed of two-argument logic products and sums. It enables us to use two-input gates performing logic AND and OR functions. It means that we can effectively and directly implement a GS on the base of FPGAs with fine-grain architectures, such as XC6200 family [Xili97, LySt96, Maye97]. The method of direct transformation of GSs to circuits with gates was presented, for example, in [Bara94].

If we consider any HGS Γ_v for which $Z(\Gamma_v) \cup \Theta(\Gamma_v) = \varnothing$ (see section 7.3), we can apply known methods of logic synthesis [Bara94, BaSk86, Skl84a] to design a scheme that implements the given behavior. The problem is how to perform switching to various levels? This problem can be efficiently resolved using a hierarchical FSM (HFSM) with stack memory [Skl84a, Skl84b]. At the beginning, the top of the stack is the register, which is used as the HFSM memory for the HGS Γ_1. Suppose it is necessary to perform an algorithm for a component ε_v of Γ_1 and $\varepsilon_v \in Z(\Gamma_1) \cup \Theta(\Gamma_1)$. In this case we can increment the stack pointer by activating y^+ and set the new register that is now located on the top of the stack, into the first state for Γ_v. As a result, the previous top of the stack stores the interrupted state of Γ_1, and the new top of the stack acquires the state of the entry point for Γ_v. The same sequence of steps can be applied to other levels. When the execution of the HGS Γ_v is being terminated, we will perform the reverse sequence of steps in order to return back to the interrupted HGS. In this case we decrement the stack pointer by activating the output y^-.

Now let us consider the physical scheme that is chosen as the base for future synthesis (see figure 7.3). The stack memory is used to keep track of the calls of macrooperations (logic functions) in the HGS Γ_1. The block S_2 has an input register which stores values of state variables $\tau_1,...,\tau_R$. The values of $\tau_1,...,\tau_R$ are converted to output signals $y_1,...,y_N$ (we are assuming that the Moore model is considered). In case of the call of a new HGS the values $\tau_1,...,\tau_R$ are converted to the code of entry point (to the code of entry state) of the new HGS, which appears on the outputs $MD_1,...,MD_R$.

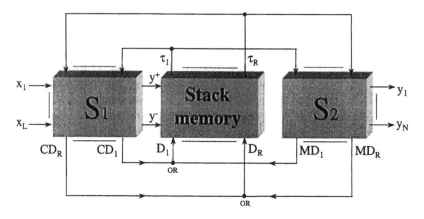

Figure 7.3. Structural scheme of a hierarchical finite state machine

If there are no transitions from one HGS to another HGS, the HFSM operates like an ordinary FSM. If it is necessary to call a new HGS Γ_v, in order to perform either a macrooperation or a logic function ε_v, the following sequence of actions will be carried out:

- the code K of a state with ε_v is stored in the selected register of the stack memory. The code K is stored in the input register of the block S_2;

- the stack pointer is incremented ($y^+=1$). As a result, a new register RG_{new} of the stack memory will be selected as the current register of our HFSM. The previous register RG_{new-1} stores the state of the HFSM when it was interrupted (in which we passed the control to the HGS Γ_v);

- the code K is converted to the code of the first state of the HGS Γ_v, which is generated on the outputs $MD_1,...,MD_R$ of the block S_2. The outputs $CD_1,...,CD_R$ of the block S_1 will be set to 0. Now the HGS Γ_v is responsible for control from this point until it is terminated;

- after termination of Γ_v, the microoperation $y^-=1$ is generated in order to return to the interrupted state. As a result, control is passed to the state in which we called the HGS Γ_v. The subsequent actions are performed in the same manner as for an ordinary FSM.

There are two states a_0 and a_1 in HFSM that are assigned predefined values: a_0 is the initial state of HGS Γ_1 and a_1 is a returned state. We will only consider the Moore HFSM below (a similar approach can be developed for Mealy HFSM). For the Moore HFSM, the output function is implemented in the block S_2 (see figure 7.3). Compared to traditional FSM, HFSM has a more complicated sub-sequence of events, which affect transitions. A correct sub-sequence is achieved by a specific synchronization [SkR96a].

Synthesis of the HFSM involves the following steps: Step 1: converting a given set of HGSs to a state transition table. Step 2: state encoding. Step 3: combinational logic optimization and design of the final scheme.

The first step is divided into three sub-steps that are: 1) marking the HGSs with labels (or with states); 2) recording all transitions between the states in the extended state transition table; 3) transforming the extended table to an ordinary form.

In order to mark the given HGSs (see sub step 1) it is necessary to perform the following actions:

- the label a_0 is assigned to the node *Begin* and to the node *End* of the main HGS (Γ_1);

- the label a_1 is assigned to all nodes *End* of the HGSs $\Gamma_2,...,\Gamma_V$;

- the labels $a_2,a_3,...,a_M$ are assigned to the following nodes and inputs: 1) the rectangular nodes in HGSs $\Gamma_1,...,\Gamma_V$; 2) the inputs of rhomboidal nodes, which contain logic functions, and which directly follow either other rhomboidal nodes, or rectangular nodes with macrooperations; 3) the *Begin* nodes of the HGS $\Gamma_2,...,\Gamma_V$ which have connections to rhomboidal nodes with logic conditions; 4) the input to a rhomboidal node with a logic condition that follows a direct connection from the *Begin* node of the main HGS (Γ_1);

- apart from a_1, all the labels in the various HGSs must be unique;

- if any node (input) has already been labeled, it must not be labeled again.

After applying this procedure to the HGSs shown in figure 7.2, we have obtained the HGSs labeled with $a_0,..,a_{20}$ (M=21). In order to build the extended state transition table (see sub step 2), it is necessary to perform the actions listed below:

- record all transitions $a_m X(a_m, a_s) a_s$ for the block S_1, where $a_m \in \{a_2,...,a_M\}$, $a_s \in \{a_0,...,a_M\}$;

- record all transitions $a_0 a_s$, $a(\varepsilon_v) a_k$ for the block S_2, where a_s is the label for a node (for an input) which follows directly from the **Begin** node of the HGS Γ_1, $a(\varepsilon_v)$ is a node (input of a node) containing the element ε_v, which is either a macrooperation or a logic function, a_k is the label for a node, which starts the HGS for the element ε_v (this is an entry point of the HGS);

- record all sets $YZ(a_m)=Y(a_m) \cup Z(a_m)$, which are subsets of the microoperations $Y(a_m)$ and the macrooperations $Z(a_m)$ generated in the rectangular node marked with the label a_m. It is acceptable that $YZ(a_m) \cap Z=\varnothing$, $YZ(a_m) \cap Y=\varnothing$.

In order to convert the table from the extended form to an ordinary form (see sub-step 3), it is necessary to perform the following actions:

- replace each macrooperation with the new output variable y^+;

- generate y^- in the state a_1;

- suppose a_u is the input label of a rhomboidal node containing the logic function $\theta_i = \varepsilon_v$. In such a case we have to add y^+ to the set $Y(a_u)$. If the input of the rhomboidal node containing θ_i has not been labeled, we have to add y^+ to the sets $Y(a_{f1}), Y(a_{f2}),...$ where the states $a_{f1}, a_{f2},...$ have been assigned to nodes from which follow direct arrow lines to the considered rhomboidal node θ_i;

- in order to save the calculated value of a logic function, and to test this value in conditional transitions, we use the technique described in [Skl84a].

The new table is the ordinary state transition table [Bara94] and we can apply a variety of known methods [Bara94, BaSk86, Skl84a, Skly96] to carry out steps 2 and 3.

After applying the first step to the HGSs in figure 7.2, we will built the table below. This combines extended and ordinary forms (all transitions implemented in the scheme S_2 are shaded; signals y^+ and y^- can be generated in any scheme). Macrooperations and logic functions are included only into the extended form, and have to be replaced with symbols in square brackets. The prefix n denotes an inverted value of the respective variable x_i.

$a_m \{Y(a_m)\}$	a_s	$X(a_m, a_s)$
a_0	a_2	1
$a_1 [y^-]$		1
$a_2 \{y_1, y_3\}$	a_4	$x_6 x_7$
	a_6	$nx_6 x_7$
	a_3	nx_7
$a_3 \{y_2, y_5\}$	a_4	1
a_4	a_{12}	1
$a_4 \{z_3\} [y^+]$	a_5	1
$a_5 \{y_1\}$	a_0	1
a_6	a_7	1
$a_6 \{y_3, z_2\} [y^+]$	a_5	x_3
	a_6	nx_3
$a_7 \{y_4, y_6\}$	a_{11}	$x_4 x_5 x_6 x_7$
	a_7	$x_4 x_5 x_6 nx_7$
	a_{10}	$x_4 x_5 nx_6$
	a_9	$x_4 nx_5$
	a_8	nx_4
a_8	a_{12}	1
$a_8 \{y_7, z_4\} [y^+]$	a_9	1
$a_9 \{y_8\}$	a_{10}	1
$a_{10} \{y_7, y_8\}$	a_{11}	1
$a_{11} \{y_1\}$	a_1	1
$a_{12} \{y_9, y_{10}, y_{11}\}$	a_{15}	$x_3 x_8$
	a_{14}	$x_3 nx_8$
	a_{16}	$nx_3 x_8$
	a_{13}	$nx_3 nx_8$
$a_{13} \{y_{10}\}$	a_{16}	1
$a_{14} \{y_{11}\}$	a_{16}	1
$a_{15} \{y_9, y_{11}\}$	a_{16}	1
a_{16}	a_{18}	1
$a_{16} \{z_4\} [y^+]$	a_{17}	1
$a_{17} \{y_1\}$	a_1	1
$a_{18} \{y_{12}\}$	a_{18}	x_1
	a_{19}	nx_1
$a_{19} \{y_{13}\}$	a_{19}	x_2
	a_{20}	nx_2
$a_{20} \{y_{14}\}$	a_1	1

Table 7.1. Extended and ordinary state transition table

7.5 Structured Design of Reconfigurable Devices Based on Predefined Schemes

The method considered in the previous section enables us to carry out formal synthesis of the control unit from the given hierarchical description. The combinational part of the scheme is composed of REs, which calculate values of Boolean functions that cause various transitions. In general each transition might need a few (more than one) REs to be activated [BaSk86, Skl84a], which makes it difficult to provide extensibility and flexibility (change-ability) for the designed scheme. For many practical applications, if we want to modify the behavior of the control unit, we have to repeat the synthesis from the beginning.

Extensibility and flexibility can be achieved with virtual control circuits, which are able to replace virtual components of a control algorithm, such as virtual HGSs, and to modify them if necessary. In order to simplify the replacement of different components we can provide direct association between HGSs and REs. In this case if we want to modify/change the HGS we have to reprogram the respective RE.

The second way to provide the facilities mentioned above is to consider a ROM-based [BaSk86, Skl84a] (or RAM-based) combinational scheme in the assumption that such elements (ROM or RAM) can be easily reprogrammed.

The considered approaches are realized with multi-level predefined schemes [Skl84a, Skly96] that are based on the one-level scheme [Bara94, BaSk86]. The common structure of the multi-level scheme is given in figure 7.4 [Skly96]. Comparing with the scheme in figure 7.3 there are three additional blocks in figure 7.4 that are a coder (C), a selector (S) and a decoder (D). These blocks can be either included into the final structure or not, depending on the approach which we are going to exploit for logic synthesis [Skl84a, Skly96].

The first approach implies microinstruction encoding [Skl82a]. In this case we will use only one additional block D which can be constructed from either PLDs or standard decoders. The connections b, c, d will be eliminated from figure 7.4 and the final scheme will be composed of the one-level sub scheme and the block D with the connections e.

The second approach is aimed at structural table lines encoding [Skl84a]. In this case we will also use only block D, which can be based on PLD (ROM in particular). The connections a, c, d will be deleted from figure 7.4 and the final scheme will be composed of the one-level sub scheme and the block D with the connections b and e. It should be mentioned that block D is especially efficient for Mealy machines [Skl84a].

The third approach enables us to perform replacement of input variables from the set $X_0 \in \{x_1,...,x_N\}$ with new variables $p_1,...,p_G$, where $G<<N$. In this case we will use only one additional block C, which can be assembled from either PLD [BaSk86] or standard multiplexers [Skl82b]. The connections b, d, e will be eliminated from

figure 7.4 and the final scheme will be composed of the one-level sub scheme and the block C with the connections a and c.

The fourth approach is related to the use of mutually exclusive elements in the scheme of the first level [Skl82b]. In this case we will employ one additional block S. The connections b, c, e will be eliminated from figure 7.4 and the final scheme will be composed of the one-level sub-scheme with the connections a and S with the connections d.

Consider non-intersecting subsets $A_1,...,A_T$ and $A_1\cup...\cup A_T=A$. Let us perform separate state encoding in each subset A_t. Usually the length of the code is less than in case of state encoding in the total set A. Consider the set $\eta=\{\gamma_1,...,\gamma_T\}$ of variables such that $\gamma_t=1$ if and only if the control unit is in the state $a_m\in A_t$, and $\gamma_t=0$ in the opposite case (t=1,...,T). As a result, element$_t$ knows a real state from the set A by analyzing both the state from A_t and γ_t (see figure 7.4). In this case the number of inputs for each element$_t$ of the one-level scheme is equal to $]\log_2|A_t|[+1$ and often less than R (especially for complex control units implementing hierarchical algorithms). The methods of synthesis of such schemes were considered in [Skl84a, Skl82b]. The element F in figure 7.4 allows us to reduce the total number of lines (U<R, f<R). For some schemes this element denotes just special connections, for instance, some lines $\tau_1,...,\tau_R$ are connected to the combinational scheme and another part to either the Coder or the Selector.

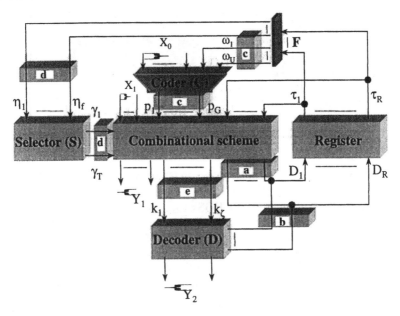

Figure 7.4. Common structure of the multilevel scheme

Actually the fourth approach enables us to provide an association between HGSs in given set and mutually exclusive elements in the combinational scheme shown in figure 7.4. It means that we can achieve what we want. The third approach makes it possible to construct the control circuit from well reprogrammable components, such as ROM (RAM) and reprogrammable multiplexers. The first and the second approaches can be combined with the third and the fourth approach and allow optimizing control units for algorithms with large number of microoperations and whose model is the Mealy FSM.

Since the fourth scheme enables us to provide direct implementation of HGSs and can be seen as a good candidate for reprogrammable and reconfigurable control circuits let us consider it in more detail (see figure 7.5).

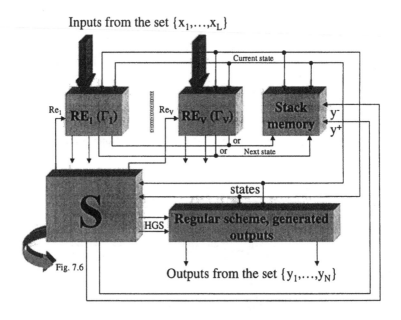

Figure 7.5. Control circuit based on mutually exclusive elements

Suppose we have a given set of HGSs $\Gamma_1,...,\Gamma_V$. Each of them is usually very simple and can be implemented with trivial control circuits. Suppose we have obtained a one to one association between the set $\Gamma_1,...,\Gamma_V$ and the set of elements $RE_1,...,RE_V$. The stack memory is used in order to keep track of the states interrupted in hierarchical sequence of HGSs to be called. The selector S is also based on stack memory (see figure 7.6) and it enables us to keep track of the REs to be activated in accordance with hierarchical sequence of the HGSs to be called.

Let us consider how to synthesize the control circuit, which implements the behavior described by the set of HGSs in figure 7.2. In this case the HGSs from a given set have to be labeled separately [Skl82b] (see labels $b_0,...,b_7$ for Moore model in figure 7.2). Now each HGS Γ_v can be implemented with autonomous circuits, such as RE_v.

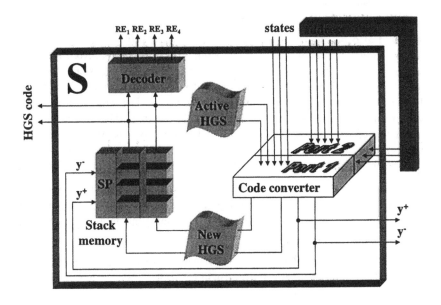

Figure 7.6. Feasible structure of the selector

The code converter is a combinational scheme, which is based on ROM (RAM) and has two ports (see figure 7.6). The first port is used in order to choose the next active HGS and the next RE respectively. The second port enables us to replace complex operations (macrooperations and logic functions), to delete them and to add new complex operations.

In order to perform the first action we have to either change data stored in proper address (see the port 2 in figure 7.6) or reprogram (reconfigure) the respective RE. In order to carry out the second action we have to reset y^+ written on respective address. In order to handle the last action we have to add a new RE and change data stored in the proper address.

The considered scheme has the following advantages. It is:

- flexible in the sense that the functions of each element can be easily modified because it implements autonomous and simple transitions for associated HGS;

- extensible in the sense that the functions of each RE can be completely changed. Besides we can extend the scheme (add new REs) without modifying the structure;

- virtual in the sense that we are able to implement a control algorithm in a scheme with restricted resources, even if the complexity of the scheme is insufficient for implementing of the entire algorithm. It is important that all links of REs are known (each HGS is a relatively independent component). So, we can replace any HGS just by freezing a single link, which is the line $\gamma_i=Re_i$ from the selector associated with the replacing RE (see figures 7.4 and 7.5). All other parts of the algorithm will not be suspended.

It should also be mentioned that we are capable of providing the execution of macrooperations in parallel using the technique [Skly87, SkR96a, RoSF97]. In this case the memory of a FSM is composed of J parallel registers, which are sequentially scanned by activating the respective sub clocks $T_1,...,T_J$. The clock cycle of the control circuit is divided into J sub clocks. The value of J defines the maximum number of macrooperations implementing in parallel. Each sub clock affects a respective register at the same level and changes its state. Suppose it is necessary to perform the macro instruction Z_i, which is composed of more than one macro operation. In this case the following sequence of actions will take place:

- each following clock cycle causes J sequential transitions in various active HGSs. If a HGS is passive the respective transition does not change its idle state. After the clock cycle will be ended all transitions in all necessary HGSs will be sequentially performed;

- if any macrooperation has been ended it waits until the last macrooperation to be included into the same macroinstruction will be completely terminated (it is supported by a special logic);

- after all macrooperations in the macroinstruction have been terminated the combinational scheme continues to run the following branches of the GS.

All necessary details related to parallel execution of macrooperations with working example were presented in [Skly87]. If we want to manage both hierarchy and parallelism we have to build the stack memory with a set of parallel registers in each level [SkR96a, RoSF97].

The basic schemes of HFSM were described in VHDL and have been carefully tested for different control algorithms (we used V-System and Synopsys VHDL analyzers) [SkR96b, RoS97a, RoS97b]. The results of our tests have shown that the target requirements of the HFSM such as extensibility, flexibility and reuse have been satisfied. All the necessary modifications and additions have been achieved through trivial reprogramming. The results of simulation for different models of HFSM were presented in [SkR96b, RoS97a, RoS97b], where different practical methods of synchronization were also suggested.

The experiments were organised as follows. We have designed a C++ program, which provides an automatic conversion of any given set of HGSs to a structural table that is similar to the table 7.1. After that we can apply a variety of known methods in order to synthesise the final logic scheme, which implements the given behavior (the given set of HGSs). The models, to be described in VHDL, allow providing required experiments with behavioral specification, such as HGSs, and with structural tables. The results of the C++ program can be embedded in VHDL models by modifying special tables, which were incorporated in VHDL code. As a result we have been able to test the correctness of the conversion. On the other hand we can provide different modifications of the initially specified HGSs in order to prove experimentally that they can be easily mapped on to the resulting scheme that has been already implemented in hardware. This work has been done and the results have shown that all the requirements, such as extensibility, flexibility and reuse, have been satisfied. The primary structures of modifiable schemes (see figures 7.3 and 7.5) were also described in VHDL and carefully examined. Since the connections between components in structural VHDL models are regular, the models have been easily extended and modified (see also papers [RoS97a, RoS97b], which exhibit different VHDL models and waveforms demonstrating the results of simulation). The property of regularity provides direct support for trivial reconfiguration, which nevertheless allows implementing all required modifications in control algorithms. Primary components of control circuits considered in the chapter might be implemented either in a gate-base or a matrix-base fashion using the methods [Skl84a, BaSk86, etc].

7.6 Using Object-Oriented Techniques

The models of digital circuits can be visualized at different levels of abstraction. In this chapter we are interested in the architectural level, which is characterized by a set of operations and in the logic level, where a circuit evaluates and calculates a set of logic functions [Mich94].

Architectural synthesis (which is also called high-level synthesis or structural synthesis) enables us to generate a macroscopic structure of the circuit represented by a set of blocks. In other words, the synthesis has to define a datapath as a set of interconnected resources (such as registers and functional blocks), and a logic model of a control unit [Mich94]. The principal tasks of the architectural level synthesis and optimization are scheduling, resource sharing, binding and datapath and control generation (these are described in detail in [Mich94]).

Logic level synthesis and optimization techniques are applied to combinational and sequential circuits and allow us to obtain a physical implementation of the scheme as a netlist of gates or other elements, such as logic, matrix or functional cells [Bara94, BaSk86, Skl84a].

The composition of a datapath and a control unit is an operational device, which might also be viewed as a hardware object (see figure 7.7). Actually digital systems can be modeled as a set of communicating hardware objects. We can ask the hardware objects to perform what they do by sending them messages [Booc94]. Each message affects the respective control unit and forces it to behave in accordance with what we ask. In this case the system is being decomposed in such a way that its components are key abstractions in the problem domain. This kind of decomposition, which is based on a collection of communicating objects aimed at achieving some desired functionality, is called an object-oriented decomposition (OOD) [Booc94].

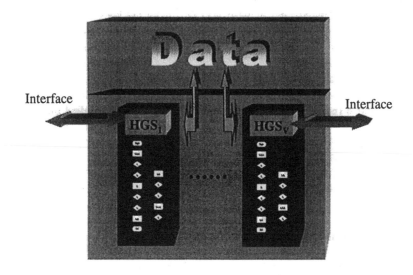

Figure 7.7. Hardware object.

It should be mentioned that the desired behavior of an individual hardware object is provided by its control unit, which is also responsible for an external interface. In other words the control unit responds to incoming (external) messages by performing things that the messages ask for.

Let us set a correspondence between messages and macrooperations described by the respective HGSs. Indeed, any message asks for the execution of the desired set of predefined actions and the associated macrooperation performs them by calling the respective sub-set of HGSs. Finally all acceptable messages to the hardware object, which specify its interface, can be associated with the set of publicly accessible HGSs describing corresponding macrooperations (see figure 7.7).

The set of HGSs can be implemented in control circuits using the considered above approach based on the model of HFSM. Since HFSMs possess the properties of extensibility, flexibility, virtualability and HGSs might be reused, we can provide the same characteristics for the respective hardware objects. These objects might be efficiently and easily implemented in dynamically reconfigurable virtual circuits, such as FPGAs [Xili97, LySt96, Maye97]. It also enables us to create a library of reusable HGSs, which facilitate the design of similar circuits [Mart95], and the library of reusable structural blocks, implementing reusable HGSs (see previous section).

Finally each hardware object has been specified by a set of HGSs with attached local datapath (see figure 7.7) synthesized by applying known technique [Mich94].

On the other hand in order to build digital systems we can use known methods of OOD [Booc94] and such ideas of object-oriented programming (OOP) as encapsulation, inheritance, polymorphism, run-time support, etc. Since digital systems to be designed according to this approach are evolved incrementally from smaller systems in which we have confidence, the risk of incorrect final results is greatly reduced [Booc94].

The primary purpose of encapsulation in OOP is the separation of the interface from implementation. The interface of a hardware object provides its outside view and therefore emphasizes the abstraction while hiding its structure. By contrast, the implementation of a hardware object is its inside view, which encompasses the secrets of its behavior (see figure 7.7).

The interface can be divided into accessible, partially accessible and inaccessible parts [Skly96] (compare this with public, protected and private attributes in OOP). For example, such public data of object **O** as a stack hardware object, which belongs to **O**, can be used in order to provide information exchange between **O** and other objects. On the other hand the stack object providing such facilities could be reused for many different objects. All private members (data and HGSs) of an object are unreachable from outside, i.e. they are secured from non authorized access.

Inheritance in OOP is a process of building a new class (derived class) from an existing class (base class). When we build a derived class we want to inherit useful properties from its base class and extend or modify the functionality of the derived class. The latter could be done by: adding new data members; adding new member functions (adding new HGSs in our case); overriding (redefining) inherited member functions. It is very important that inheritance enables us to provide reuse of an existing code and to produce derived classes, which are more powerful than their base classes.

When we are building a hierarchy in OOD we are dealing with inheritance, which can be considered as a basic way to represent multilevel abstractions. The purpose of inheritance is to provide a commonality of representation and a calling interface. The low-level mechanisms of managing inheritance were considered in [Skly97] and they can be directly adapted to hardware objects. As a result we can reuse the configuration information of reconfigurable hardware from the base object in order to configure the hardware for the respective derived object.

Generally speaking, reuse denotes the ability of a device to be used again. Sometimes we want to add functionality or to change behavior. In the approach we are considering, we do not need to start the design process again from the beginning. The new circuit inherits the invariable part of the previous interface and just adds (or replaces) the part that is different in the new context. This is achieved through inheritance mechanisms combined with the concepts of polymorphism and run-time support that are provided by virtual elements.

Actually virtual elements (macrooperations and logic functions) provide a mechanism for run-time binding (we will also call it as late or dynamic binding), which makes it possible the redefinition (overloading) problem during control, rather than at the process of synthesis or configuring (we will call this process as static binding) [SkRF97]. Static binding is applied when we carry out steps of either logic synthesis or configuring. Dynamic binding allows us to cope with the problem of HGS redefinition during physical control (during run-time). It gives us the following advantages:

- we can improve the previous version of macrooperation (logic function) given earlier (for instance we can replace an old version with the new version);

- we can provide run-time fitting. In other words during control we can compute and choose the best sub-algorithm (HGS) for given conditions, which is known only at execution time.

In certain circumstances we can combine control and synthesis in order to search either for the best sub-algorithms or for optimal implementations.

The primary benefit and distinguishing feature of the HFSMs is that they allow us to set up and to perform all necessary operations with relatively independent sub-algorithms, described by the respective HGSs. This gives us the following advantages:

- the ability to reuse previously constructed HGSs and previously designed structural blocks of HFSMs. By investing a little extra effort in the design, we can create a library of reusable components such as HGSs, which will facilitate the design of similar circuits. This approach is very similar to the methods of object-oriented analysis and design [Mart95]. The basic reusable component is a separate HGS, which can be designed and tested in reconfigurable hardware (using the technique of fast prototyping, for example) independently on the rest of the control algorithm. In [Mart95] there is an example of the design of reusable components for a FSM;

- the flexibility in the control algorithm in terms of possible trivial re-switching between relatively autonomous and simple components such as HGSs and possible modifications of HGSs. This can be done because the scheme itself becomes more flexible. If we want to modify or to change a component, we do not need to redesign the scheme from the beginning. We just have to redesign a single component;

- the extension of a given control algorithm becomes a relatively simple matter. Indeed we can easily solve the problem of extending the behavior of a HGS Γ_v, by modifying it. On the other hand we can add new HGSs without changing the basic frame of control circuit;

- the control circuit becomes virtual. It means that logic resources of the circuit might be smaller than we actually need in order to implement the given control algorithm. The scheme can be modified during run-time in order to use the same hardware (the same REs) for implementing various HGSs. All HGSs, which we need, are stored in an external (supplementary) memory in form of configuring information for the respective hardware. This information can be further used for loading of static (non virtual) HGSs and loading and replacing of dynamic (virtual) HGSs;

- an incomplete top-down design of the control unit based on virtual and pure virtual macrooperations (logic functions) and a simplification of the debugging process.

Generally speaking any dynamically reconfigurable device is a virtual circuit, which denotes that the same physical scheme can be used to implement different logic nets. The idea is to combine a flexible and extensible behavioral description, i.e. HGSs, with virtual circuits being used to implement this description. In order to model hardware objects whose behavior have been described by a set of HGSs on the base of virtual circuits, we can apply the technique of OOP in general and OOD in particular [Booc94, Mart95].

On the other hand the internal structure of hardware objects containing local datapath and functions (control circuits) can be carefully optimized by involving methods of digital design. Using HGSs to describe functions of hardware objects enables us to shift problems of optimization of digital circuits from the traditional way [Mich94] to the optimizing of HGSs [Bara94].

References

[AtSi93] P. M. Athands, H. F. Silverman, *"Processor Reconfiguration trough Instruction-Set Metamorphosis"*, Computer, vol. 26, n° 3, pp. 11-17, March, 1993.

[Bara94] S. Baranov, *"Logic Synthesis for Control Automata"*. Kluwer Academic Publishers, 1994.

[BaSk86] S. Baranov, V. Sklyarov, *"Digital Devices Based on Programmable Matrix LSI"*. Radio and Communications, Moscow, 1986.

[Booc94] Grady Booch, *"Object-Oriented Analysis and Design"*, Second Edition, The Benjamin/Cummings Publishing Company, Inc., 1994.

[Bost96] G.Bostock, *"FPGAs and Programmable LSI. A designer's handbook"*. Butterworth-Heinemann, 1996.

[Brow96] S.Brown, *"FPGA Architectural Research: A Survey"*, IEEE Design & Test of Computers, Vol. 13, No 4, pp 9-15, 1996.

[ChLe96] D.Chrepacha, D.Lewis, *"DP-FPGA: An FPGA Architecture Optimized for Datapath"*, VLSI Design, Vol. 4, No 4, pp. 329-343, 1996.

[ChLR93] Colin Charlton, Paul Leng, Mark Rivers, "An Object-Oriented Model of Design Evolution", *Microprocessing and Microprogramming*, vol. 38, Numbers 1-5, pp. 441-448, September, 1993.

[ELLV97] Stephen Edwards, Luciano Lavagno, Edward A.Lee, Alberto Sangiovanny-Vincentelli, *"Design of Embedded Systems: Formal Models, Validation, and Synthesis"*, Proceeding of the IEEE, vol. 85, no. 3, pp. 366-390, March, 1997.

[Gokh91] M. Gokhale et al, *"Building and Using a Highly-Parallel Programmable Logic Array"*, Computer, vol. 24, n° 1, pp. 81-89, January 1991.

[Hura97] Laszio Huray, "Interoperable Objects for Distributed Real-Time Systems", *Embedded System Programming Europe*, pp. 9-22, May, 1997.

[IsSa93] C. Iseli, E. Sanchez, "Spyder: a Reconfigurable VLIW Processor Using FPGAs", *Proc. of the IEEE Workshop on FPGAs for Custom Computing Machines*, pp. 17-24, April 1993.

[Katz94] Randy H. Katz, *"Contemporary Logic Design"*, The Benjamin/Cummings Publishing Company, Inc., 1994.

[KiSk79] Victor Kirpichnikov, Valery Sklyarov, *"Description and Synthesis of Control Devices"*. USSR Academy of Science, Technical Cybernetics, N 1, pp 127-137, Moscow, 1979.

[Kuma94] S.Kumar, J.H.Aylor, B.W.Johnson, W.A.Wulf, *"Object-Oriented Techniques in Hardware Design"*, Computer, vol. 27, no. 6, pp. 64-70, June, 1994.

[LySt96] Patric Lysaght, Jon Stockwood, *"A Simulation Tool for Dynamically Reconfigurable Field Programmable Gate Arrays"*, IEEE Trans. On VLSI Syst., vol. 4, no. 3, pp. 381-390, September, 1996.

[Mart95] Robert C. Martin, *"Designing Object-Oriented C++ Applications Using the Booch Method"*, Prentice Hall, 1995.

[Maxf96] Maxfield, Clive "Max", *"Field Programmable Devices"*, EDN, pp. 201-206, October 10, 1996.

[Maye97] John H.Mayer, *"Reconfigurable Computing Redefines Design Flexibility"*, Computer Design, pp. 49-52, February, 1997.

[Mich94] Giovanni De Micheli. *"Synthesis and Optimization of Digital Circuits"*, McGraw-Hill, Inc., 1994.

[Migu97] Giovanni de Micheli, Rajesh K. Gurta, *"Hardware/Software Co-Design"*, Proceeding of the IEEE, vol. 85, no. 3, pp. 349-365, March, 1997.

[Pirp97] Eric Pirpich, *"Designing a more Flexible Programmable Logic Device - the XPLA"*, Electronic Engineering, pp. 65-70, January, 1997.

[RoS97a] António Adrego da Rocha, Valery Sklyarov, "VHDL Modeling of Hierarchical Finite State Machines", *Proceedings of the Fifth BELSIGN Workshop*, Dresden, Germany, April, 1997.

[RoS97b] António Adrego da Rocha, Valery Sklyarov, *"Simulação em VHDL de Máquinas de Estados Finitas Hierárquicas"*, Electrónica e Telecomunicações, Vol 2, N 1, pp. 83-94, 1997.

[RoSF97] Antonio Adrego da Rocha, Valery Sklyarov, Antonio de Brito Ferrari. "Hierarchical Description and Design of Control Circuits Based on Reconfigurable and Reprogrammable Elements", *Proceeding of the International Workshop on Logic and Architectural Synthesis - IWLAS'97*, Grenoble, December, 1997.

[Skl82a] Valery Sklyarov, "Using Decoders in Control Units", *University News. Instruments*, N 12, pp. 27-31, Leningrad (St.Petersburg), 1982.

[Skl82b] Valery Sklyarov, "Synthesis of Control Units Based on Programmable Logic Devices", USSR Academy of Science, *Technical Cybernetics*, N 5, pp. 59-69, Moscow, 1982.

[Skl84a] V. Sklyarov, "Synthesis of Finite State Machines Based on Matrix LSI", *Science and Technique*, Minsk, 1984.

[Skl84b] Valery Sklyarov, "Hierarchical Graph-Schemes", Latvian Academy of Science, *Automatics and Computers*, no. 2, pp. 82-87, Riga, 1984.

[Skly87] Valery Sklyarov, "Parallel Graph-Schemes and Finite State Machines Synthesis", Latvian Academy of Science, *Automatics and Computers*, no. 5, pp. 68-76, Riga, 1987.

[Skly96] Valery Sklyarov, *"Applying Finite State Machine Theory and Object-Oriented Programming to the Logical Synthesis of Control Devices"*. Electrónica e Telecomunicações, 1996, Vol 1, N 6, pp. 515-529.

[Skly97] Valery Sklyarov, *"Understanding and Low Level Implementation Basic OOP Constructions"*, *Electrónica e Telecomunicações*, Vol 1, N 7, pp. 729-738, 1997.

[SkR96a] Valery Sklyarov, António Adrego da Rocha, "Sintese de Unidades de Controlo Descritas por Grafos dum Esquema Hierarquicos", *Electrónica e Telecomunicações*, vol. 1, no. 6, pp. 577-588, 1996.

[SkR96b] Valery Sklyarov, António Adrego da Rocha, "Synthesis of Control Units Described by Hierarchical Graph-Schemes", *Proceedings of the Fourth BELSIGN Workshop*, Santander, Spain, November, 1996.

[SkRF97] Valery Sklyarov, Antonio Adrego da Rocha, Antonio de Brito Ferrari. "Applying Procedural and Object-Oriented Decomposition to the Logical Synthesis of Digital Devices", *Proceeding of the Second International Conference on Computer-Aided Design of Discrete Devices - CAD DD'97*, Minsk, November, 1997.

[VBRS96] Jean E.Vuillemin, Patrice Bertin, Didier Roncin, Mark Shand, Hervé H.Touati, Philippe Boucard, *"Programmable Active Memories: Reconfigurable Systems Come of Age"*, IEEE Trans. On VLSI Syst., vol. 4, no. 1, pp. 56-69, March, 1996.

[WRVr95] S.Wilton, J.Rose, Z.Vranesic, "Architecture of Centralized Field-Configurable Memory", *Proc. Third ACM Int'l Symp. On Field Programmable Gate Arrays*, pp. 97-103, Assoc. For Computing Machinery, New York, 1995.

[Xili97] Xilinx, "XC6200 Field Programmable Gate Arrays", *Xilinx Product Description (Version 1.10)*, April 24, 1997.

[ZhPa96] Wei Zhao, Christos A. Papachristou, "Synthesis of Reusable DSP Cores Based on Multiple Behaviour", *IEEE/ACM International Conference on Computer Aided Design*, San Jose, California, pp. 103-108, November 10-14, 1996.

8 ESTIMATION OF CIRCUIT PHYSICAL FEATURES IN HIGH-LEVEL SYNTHESIS BASED ON STANDARD CELLS

Milagros Fernández
Hortensia Mecha

8.1 Introduction

High Level Synthesis (HLS) or even higher abstraction levels of synthesis are needed in order to cope with increases in design complexity due to recent advances in semiconductor technology and the urgency to reduce the time-to-market of products. Estimations at physical level are of paramount importance in a HLS tool, because they let you know whether the design meets the constraints before implementing it, they make it easier to take the right decision at earlier design tasks and when the constraints are not met they help to identify the design level which should be modified. Moreover, these estimations help to avoid the time consuming steps of placement and routing that would be run within each iteration of the design process to ensure that restrictions are met.

The estimations presented in this paper are related to a standard cells design style, since it is a main representative of custom design. Estimations of area and delay of an integrated circuit, IC, are studied, in order to help the HLS tool. The low-level tool connected with it has been CADENCE.

Estimations are particularized on interconnections due to the fact that have been slightly treated and, as technology improves, they are becoming responsible for a greater percentage of area and delay in ICs.

179

J.C. López et al. (eds.), Advanced Techniques for Embedded Systems Design and Test, 179-199.
© 1998 *Kluwer Academic Publishers.*

8.2 Technological Characteristics of Designs with Standard Cells

A design implemented with standard cells is a set of rows and channels between them. Rows role is to place the cells, while interconnections among them are on channels. All the rows are of the same length, L_{row}, and the same height. CADENCE has been used with two metal layers for connections. The lower one, going through the channels, is used for the horizontal interconnections (parallel to rows), while most part of the vertical connections are placed on the upper one. This fact implies that no interconnections intersect at the same point in the same metal layer.

Each channel has a variable number of parallel tracks where interconnections are arranged. Each track can accommodate several connections with the constraint that they do not overlap. Figure 8.1 shows a design with standard cells.

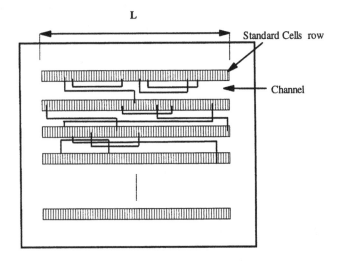

Figure 8.1. Standard cells design

Although almost all tracks have the same width, a technology parameter, there are special tracks reserved for connections among a great number of cells, (clock, power and ground), whose widths are tabulated for ranges. Placement and routing algorithms must check such a table, in order to reserve enough space for them. Interconnection width is composed of two terms, metal width plus the minimum gap between adjacent tracks that is also fixed for each technology.

8.3 Estimations of Interconnection Area for Standard Cells Designs.

Some authors have assumed that module area could easily be worked out if a module library were available [Pang88], [DeNe89]. This is only true in the case of using macrocells with a fixed structure. If the design is been implemented with standard cells, the module cells area is a constant, but when as it is usual the case, this module were connected to others, its cells would tend to be arranged according to the external influences of the other modules, arising two factors that had to be taken into account: relative cells positions change, and channels width grows. These facts are responsible for different interconnections lengths and, as a consequence, different areas, in the case that a module were isolated or interconnected to other modules.

Thus, an estimation of internal and external interconnections is needed and must be done mainly in two scenarios: First, it is necessary to take one among different options to estimate the increment of area due to the creation of each interconnection without still knowing the total number of modules and interconnections. Estimations would be only based on the type of modules that are going to be connected, on the know-how of the algorithms of placement and routing, and on the design technology. Second, when the estimation of the final area is needed, to decide if the design meets the constraints instead of generating the layout.

8.3.1 Simple Area Estimation Model: Interconnection Average Length

The tool that has been used to do the estimations is CADENCE, and it has been applied to designs implemented with two metal layers, where vertical segments of an interconnection go through a different layer, metal 2. Therefore, the length of an interconnection between cells placed on different rows is reduced to the length of the horizontal segments on metal 1, and can be computed as if the two cells were placed on the same row. The model could be applied to any other number of metal layers as long as the horizontal segments routed on metal 1 are known.

The first model introduced in this section is a simple one that considers all the connections of the same length, and equal to the average one obtained considering all the connections equally likely between any pair of cells.

Let L_{row} be the length of a row of standard cells, L_{cell} the average length of a cell, and N_{cells} the total number of cells in the row, Figure 8.2.

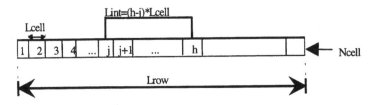

Figure 8.2. Average distance in a row

The average distance between any pair of cells is computed adding the length of the different interconnections, L_{conn_i}, and dividing by the total number of interconnections, N_{total}.

$$L_{aver} = \frac{\sum_{i=1}^{N_{total}} L_{conn_i}}{N_{total}}$$

An interconnection between cell j and cell h, h>j, will have a horizontal length given by

$$L_{conn} = (h - j) L_{cell}$$

j could take values from 1 to $N_{cells} - 1$ and h from j + 1 to N_{cells}

The result obtained for $N_{cells} \gg 1$, that it is the case for a row of standard cells, is:

$$L_{aver} = L_{cell} \frac{N_{cells}}{3}$$

As $L_{row} = L_{cell} N_{cells}$ the average length of an interconnection is given by:

$$L_{aver} = \frac{1}{3} L_{row}$$

As IC shapes implemented with standard cells are almost square, L_{row} is the side of the circuit and so L_{aver} could be computed as:

$$L_{aver} = \frac{1}{3} \sqrt{A_{total}}$$

where the total area has two terms, one corresponding to the cells area, and the other to the interconnections area. The later is computed as the area covered by the total number of interconnections in the circuit N, either internal or external to the modules. All connections have the same width, Wconn, that is characteristic of the technology used. Thus:

$$L_{aver} = \frac{1}{3} \sqrt{A_{cells} + N L_{aver} W_{conn}}$$

From this expression, we can obtain a second order equation in L_{aver} whose solution is:

$$L_{aver} = \frac{N\,W_{conn} + \sqrt{(N\,W_{conn})^2 + 4 \times 9\,A_{cells}}}{18}$$

As a consequence, module area is computed as the result of adding its cells area to the estimated area of its internal interconnections, whose number is known. The estimated area of one of these connections is given by the product of L_{aver} times W_{conn}.

In order to evaluate the quality of the predictions when using this simple model, the 5th order filter is used, a typical HLS benchmark, getting the results shown at Figure 8.3.

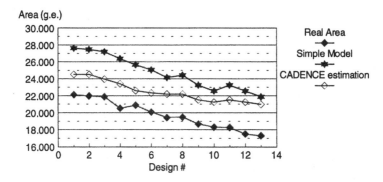

Figure 8.3. Simple Model Estimation

Area is expressed in gate-equivalent, g.e. Several designs are run, under three different scenarios. First, by using the model proposed, then with an estimation tool used by CADENCE before floorplanning, and finally, running the design till the physical implementation is done. As it can be observed, the estimated area obtained when using as interconnection length its average value, is always greater than the values obtained with CADENCE estimator. Furthermore, the estimation is not reliable, because design number 10, whose estimated area is smaller than the area of design number 11, is in fact, the opposite for the real values. Thus, it seems imperative to develop another model, that mirrors in a more efficient way interconnections.

8.3.2 Refined Area Estimation Model: Interconnections Classes

When circuit designs are studied after placement and routing, it can be observed that the modules connected together tend to be placed nearby, and so, these interconnections had a length shorter than the average value estimated with the simple model. This fact is due to the nature of algorithms of placement and routing, which have as a goal the minimization of the area of the circuit through the minimization of interconnection length. This means that the total area has been overestimated. Modules that are heavily connected are placed nearer than the modules slightly connected. Based on this idea, it could be developed a new model where there can be distinguished two different interconnection classes: close and distant interconnections. The former corresponds to exclusive interconnections between two modules that are not linked to other modules. The later takes into account connections between modules that at the same time are connected to other modules, Figure 8.4.

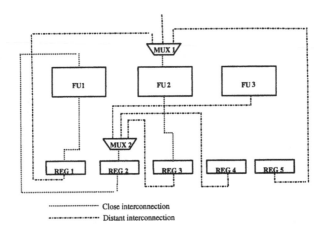

Figure 8.4. Interconnection Classes

Close interconnections can be estimated by the average length of a cell, since its connected cells are assumed to be placed together. However, distant interconnections are still being estimated by the average interconnection length.

Thus,

$$L_{close} = \overline{L_{cell}}$$

and

$$L_{distant} = L_{aver} = \frac{1}{3}\sqrt{A_{cells} + (N_{close} L_{close} + N_{distant} L_{distant})W_{conn}}$$

where N_{close} represents the number of close interconnections and $N_{distant}$ the number of distant interconnections. From this expression we can obtain a second order equation in $L_{distant}$ whose solution is:

$$L_{distant} = \frac{N_{distant} W_{conn} + \sqrt{(N_{distant} W_{conn})^2 + 4 \times 9(A_{cells} + N_{close} L_{close} W_{conn})}}{18}$$

The estimated area with this refined method that considers two classes of interconnections is:

$$Area_{estim} = A_{cells} + (N_{close}\ L_{close} + N_{distant}\ L_{distant})W_{conn}$$

If this estimation is applied to the same example of the 5th order filter, the results are similar to the results obtained with CADENCE estimator, overall for designs over of design number 9, as it is established in Figure 8.5. Furthermore, the estimation of area is reliable because one design that according to this model is better than another one, when implemented as a real circuit, is also better.

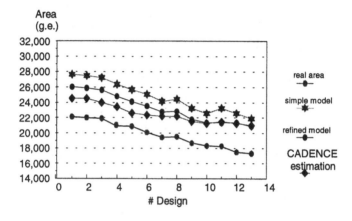

Figure 8.5. Refined Model Estimation

8.3.3 Complete Area Estimation Model: Multimodule Interconnections

The results obtained with the refined model show that the area is still overestimated, due mainly to the multimodule interconnections. This type of connections appears when the output of one module is connected to the inputs of several modules, as is the case in the example in Figure 8.6, where the output of the functional unit FU2 is linked to the inputs of three registers, REG3, REG4, and REG5. With the refined model proposed before, this multimodule interconnection tends to be treated as three point to point interconnections, and this implies that part of the interconnection is counted twice and part of it, thrice. This fact leads us to think that area estimations done by the refined model do not represent accurately the area of multimodule interconnections. If a general case is considered, when the output of a module is connected to the inputs of m modules, it is obvious that the length of the interconnection does not grow linearly with m. As m becomes larger, the increment of the total length of the interconnection is smaller, because it is possible to profit from the previous sections due to the m-1 modules formerly connected.

In order to compute the entire length of a multimodule interconnection Lmmod, an interpolation function that embodies two main requirements is used. The first gives account of the fact that as m increases, the increment of the interconnection length must decrease. The second must reflect the situation that as the number of connected modules grows, the length of the multimodule interconnection must tend to the length of the standard cells row. This fact could be explained considering Figure 8.7 that could represent the placement on standard cells of the connections of the example in Figure 8.6.

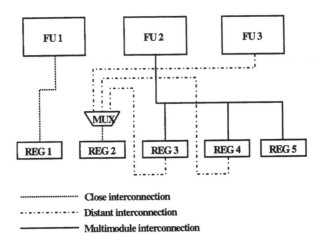

Figure 8.6. Multimodule Interconnections

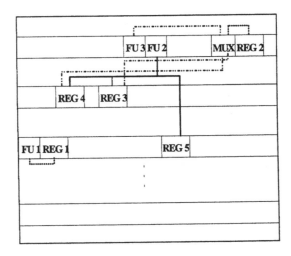

Figure 8.7. Multimodule interconnection on standard cells

The registers could be placed on different rows but, as it was explained in section 8.3.1, only horizontal tracks have to be taken into account and, the length of the multimodule interconnection could be computed as if all modules were placed on the same row. If only one module is connected, 1 in Figure 8.8, a point to point interconnection is obtained and its length would be the length of an average interconnection. If we assume that a new module is connected, it could be consider as a point to point interconnection whose source would be 1 in Figure 8.8, and whose length would correspond to the average length of the remaining length of the row.

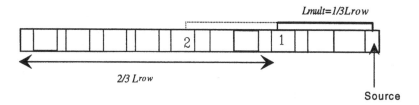

Figure 8.8. Multimodule Interconnection Length

So for m=1 the length of the interconnection is

$$L_{mmod_1} = \frac{1}{3} L_{row}$$

The length increment experimented when one second module is added is given by:

$$\Delta L_{mmod_2} = \frac{1}{3} \frac{2}{3} L_{row}$$

If a new module is added

$$\Delta L_{mmod_3} = \frac{1}{3} (\frac{2}{3})^2 L_{row}$$

The length increment experimented when an m-th module has been added is:

$$\Delta L_{mmod_m} = \frac{1}{3} (\frac{2}{3})^{m-1} L_{row}$$

The total length of an interconnection that connects one source to m modules is the addition of the m increments:

$$L_{mmod_m} = \sum_{j=1}^{m} \frac{1}{3} (\frac{2}{3})^{j-1} L_{row} = (1 - (\frac{2}{3})^m) L_{row}$$

As L_{row} is unknown because depends on the area

$$L_{row} = \sqrt{A_{estim}} = \sqrt{A_{cells} + A_{conn}}$$

But the area due to interconnections, A_{conn} , is the result of the close, distant and multimodule interconnections. To ease the computation of this area, it could be observed that L_{mmod} could be expressed in terms of $L_{distant}$ as:

$$L_{mmod_m} = 3(1-(\frac{2}{3})^m)\frac{1}{3} L_{row} = 3(1-(\frac{2}{3})^m) L_{distant}$$

In order to estimate the area, it is necessary to know how many multimodule interconnections exist and how many modules are linked to each one. The only unknown would be then   Ldistant and, based on that, it would be possible to compute the area of any multimodule interconnection and, as a consequence the complete area of the circuit. The shown method has a complexity proportional to N, where N is the number of modules in the circuit.

Figure 8.9 presents the results when the complete model is applied to the same 5th order filter example. As it can be observed, the estimation is reliable as the one obtained with the refined method but, it is also accurate because the errors are always smaller than 5%.

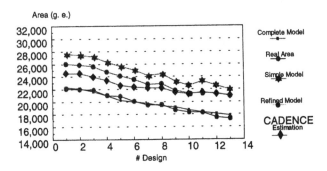

Figure 8.9. Complete Model Estimation

8.4 Area Estimations at Different Steps of a High Level Synthesis Tool

Area estimations are particularly important for a HLS tool, because they allow to know whether a design meets the area constraints before implementing it. They make it easier to take the right decision at earlier design tasks, as scheduling and hardware allocation, (**HA**), and when the constraints are not met, they help to identify the design level which should be modified.

The HLS tool used, FIDIAS[SMTH95], needs to take decisions based on estimations during several steps of the design process: after scheduling, during and after HA, and during control generation. The estimations after scheduling are needed because it is convenient to know the quality of the scheduling, regarding the designs that are going to evolve from it. Of course, HA is not done yet, therefore the area estimation at this point of the design cycle is related only with a minimum circuit area that it is possible to know, because the design has a minimum area due to registers, to functional units, (**FU**)s and to interconnections.

The minimum number of registers could be calculated using the method described in [MoHF96]. Each register is composed by a set of standard cells and its internal interconnections, whose area is unknown but, whose type and number are known. The minimum area due to FUs could be computed taking into account that a preallocation is done after scheduling, allowing to know the number and type of functional units which are going to implement the scheduled Data Flow Graph, (**DFG**). As in the case of registers, the area of standard cells is known. Regarding the internal interconnections only the type and number is available, information that will be used later to compute the interconnections area. As far as the minimum interconnection area is concerned, we have to consider the aforementioned internal modules connections, along with the external connection among modules. To compute the minimum number of these connections, the maximum number of functional units input/outputs among control steps are used. This estimation is coarse, but the information that is known at this step of the design cycle does not allow to obtain more refined estimations. Once the number and type of interconnections are known, it is possible to apply the method of section 8.3.3 to estimate their area and, as a consequence, to estimate the minimum area of the design. If this value is greater than the area restriction given by the user, it is mandatory to reject the scheduling and a new one has to be proposed, either by changing the value of cycle time or by increasing the number of control steps. This permits to diminish the parallelism and to increase the hardware reusing. If the estimated minimum area is smaller than the restriction, the next step, hardware allocation, could be addressed.

In FIDIAS estimations are not only performed on a complete design but also any time it has to be taken a decision among the whole set of alternatives, as hardware reusing or embodiment of new modules. In either of the cases, it is necessary to estimate the area of the new modules, or the area of the new interconnections in the case of hardware reuse.

As it was previously indicated, the module area has two terms, one due to standard cells and the other to interconnections. The knowledge about the later is reduced to their type and number. Close interconnections are characterized by the technology, but distant and multimodule area interconnections only could be known when the final design is known. It is also necessary to estimate the area when a new interconnection is created, thus if this interconnection is distant or multimodule the same could be said. The only estimate that it is available when the first design of a particular circuit is being created, is the estimate of the minimum area obtained during scheduling, and this is the one that it is used to estimate the area of distant and multimodule interconnections. As it is shown at Figure 8.10, area estimation of a distant interconnection at the first design is much smaller than the obtained at subsequent designs of the same circuit. This is due to the fact that the first design uses the minimum area of the circuit to estimate it, and the others use the estimation of the previous design area. It can also be observed that for the following designs, the variation is not greater than 1% in spite of the fact that the estimate of the complete area of the previous design is used to estimate the interconnection area. When HA has ended, the result obtained can be used to evaluate the quality of it, and to decide if a new task can be addressed or, it is advisable to reinitiate the process.

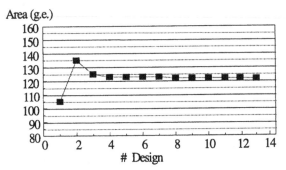

Figure 8.10. Distant interconnection area estimation

A final step in HLS is the generation of the Control Unit, (**CU**) that is an automatic process and whose size can not be optimized. Nevertheless, CU area has to be included as part of the area of the integrated circuit. FIDIAS only considers the influence of CU on area during operation scheduling, in which the goal of area optimization is translated to minimizing the number of control steps. The reason is that the number of control steps sets the size of the microinstructions ROM if it is a microprogrammed CU, or the state machine if it is a wired CU.

Estimations of module area in the CU can be done in an analogous way to estimations of module area in datapaths. The only difference can be the estimation of the ROM area. If it is implemented by a macrocell, it is easy to obtain its estimation in terms of the microinstruction width and the number of microinstructions. The size of macrocells is fixed, and its value does not change when they are placed at the final design.

8.5 Interconnection Delay Estimation

In technologies around 1μ interconnection delay is comparable to functional delay [WePa91], so it is important to estimate it before calculating the cycle time. In order to estimate the interconnection delay, we need to find a physical model to represent the interconnection. As a basis for our study we start with a simple model where interconnections are only point to point. Consequently, an interconnection is a line with one end point driven by a MOSFET and the other connected to the gate of another MOSFET, as it is shown in Figure 8.11.

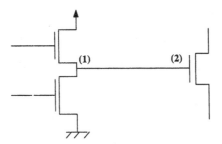

Figure 8.11. Interconnection schema

For current technologies, the resistance and capacitance of the line can not be considered zero. Although there are some different models to represent it, the best one matching the interconnection behavior is $\pi 3$, that permits to estimate the delay with less than 3% error [Saku83]. This model is represented in Figure 8.12, where R and C are the total resistance and capacitance of the wire, C_t is the capacitance of the load transistor and R_t is the equivalent resistance of the driving transistor. If the width for an interconnection is fixed, and r and c are the resistance and capacitance of a wire per length unit then: $R=r*L$ and $C=c*L$, where L is the interconnection length.

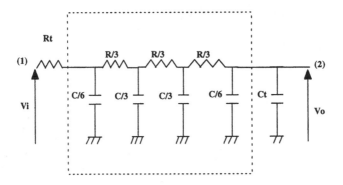

Figure 8.12. $\Pi 3$ Model

To calculate the interconnection delay we use $t_{0.9}$, that is the time delay from the moment a step voltage V_i is applied to the source point (1) until V_o at node (2) reaches the value $0.9V_i$. To compute $t_{0.9}$ it is necessary to study the circuit

$$H(s) = \frac{V_o(s)}{V_i(s)} = \frac{F}{As^4 + Bs^3 + Ds^2 + Es + F}$$

where:

$$A = C^3 R^3 R_t [C + 6C_t]$$
$$B = 6C^2 R^2 [C[R + 12 R_t] + 6C_t[R + 9 R_t]]$$
$$D = 81CR[C[4R + 19 R_t] + 2C_t[8R + 27 R_t]]$$
$$E = 4374[C[R + 2 R_t] + 2C_t[R + R_t]]$$
$$F = 8748$$

being s the complex frequency.

The transfer function denominator is a fourth order polynomial in s, so it has four roots, the poles of H(s). Their values depend on the relationship $C_t*R_t/C*R$, and therefore on the interconnection length and on the technology. The time domain response of the circuit depends on the poles position. There is a **dominant pole** whenever a pole is much lower than the others, that are at least a decade away. It can be proved [Kuo85] that the dominant pole determines the time domain response, and therefore the system can be approximated by a first order system.

$L(\mu)$	R_tC_t/RC	pole2 (rad/ps)	pole1 (rad/ps)	pole2/pole1
1	1.39E+07	-6.71E+05	-2.00E-02	3.36E+07
2	3.47E+06	-1.68E+05	-1.99E-02	8.43E+06
3	1.33E+06	-6.42E+04	-1.99E-02	3.23E+06
5	5.56E+05	-2.70E+04	-1.98E-02	1.36E+06
10	1.39E+05	-6.79E+03	-1.96E-02	3.46E+05
20	3.47E+04	-1.72E+03	-1.93E-02	8.93E+04
30	1.54E+04	-7.76E+02	-1.90E-02	4.09E+04
50	5.56E+03	-2.87E+02	-1.83E-02	1.57E+04
100	1.39E+03	-7.61E+01	-1.69E-02	4.51E+03
500	6.86E+01	-5.09E+00	-1.09E-02	4.65E+02
1000	1.39E+01	-1.33E+00	-7.00E-03	1.90E+02
2000	3.47E+00	-4.20E-01	-4.21E-03	9.99E+01
5000	5.56E-01	-8.57E-02	-1.86E-03	4.60E+01
7000	2.44E-01	-4.09E-02	-1.24E-03	3.31E+01
10000	1.39E-01	-2.45E-02	-9.22E-04	2.66E+01

Table 8.1. Two most significant poles

It is interesting to know the variation of the two most significant poles, as a function of the relation $C_t*R_t/C*R$. Taking into account this goal, we have performed a study based on typical values of C_t and R_t from ES2 Library [ES293], a standard cells library for CADENCE, and a realistic interconnection length range. We have taken as typical values for 1μ technology, $R_t=1000\Omega$, $C_t=0.05pF$,

$r=0.04\Omega/\mu$ and $c=0.09fF/\mu$. In this study L ranges from 10000μ to 1μ, that are the values of the interconnection lengths for typical SAN examples. The results are shown in Table 8.1. In all cases the dominant pole frequency is at least 25 times lower than the other poles, so it is possible to model the system as a first order one to estimate $t_{0.9}$.

In a first order system, the response in terms of time is:

$$V_o(t) = V_i(1 - e^{-\sigma t})$$

where σ is the dominant pole.

To compute $t_{0.9}$ we take $V_o(t)$ as $0.9V_i$, and so

$$t_{0.9} = \frac{\ln 10}{\sigma}$$

The computation of the poles in a fourth order system and, therefore, its transformation into a first order system, are of constant complexity. Besides, the transmission line delay estimation is also of constant complexity, when R, C, R_t and C_t are known. This solution is valid for the range of L from 10000μ to 1μ. Nevertheless, for longer interconnections, the relationship between the two most significant poles could be lower, and then this method would not be suitable.

However, these longer interconnections have excessive delays, and, consequently, the clock cycle of the circuit should be very long in order to make sure a right electrical behavior. Currently, the designer uses different methods to avoid these long interconnection delays, for example making wider interconnections to decrease the resistance, or using buffers that accelerate the signal transmission [WeEs94]. Both solutions lead to interconnection lengths where the dominant pole method can be applied.

But the interconnection delay depends on its length, and so, it is necessary to calculate the value of L we should use to compute the signal propagation delay for all the operations in the DFG. As each interconnection length is different, in next section we will estimate a value to settle securely the right circuit electrical behavior.

8.6. Clock Cycle Selection Considering Interconnection Delay in High Level Synthesis

In HLS the selection of the clock cycle is very important because it influences the overall execution time and the size of the CU of a design. Usually, a clock cycle minimizing the dead times of the FUs performing the operations of the Data Flow Graph (DFG), causes an optimal execution time. However, generally this clock cycle is too short and the number of control steps grows. Then the size of the CU could become unacceptably large.

Scheduling and HA depend on the clock cycle, so it is important to spend some time trying to obtain a good value, in order to drive the design process to a good solution. If the design constraints are not met because the clock cycle was not properly selected, it is necessary to try a new value, to schedule again the graph and to allocate the hardware. This may take too much time.

In this selection it is very important to pay attention to the interconnection delay. In technologies around one micron, the interconnections capacitances come to the same magnitude as transistors capacitances, and therefore, the signal propagation delay becomes an important factor in the total execution time of any operation.

We present a method to find the clock period considering this factor, taking into account only point to point interconnections. This technique leaves space for improvements, such as consideration of multimodule interconnections. A basic design model, including all the elements of the data-path (DP) and the signal propagation delay, is presented.

Clock cycle computation is an important task often neglected in HLS literature. The usual approach is to use a fixed clock cycle specified by the user [WaPa95], without considering its influence over the area and performance of the final circuit. This value is optimised afterwards. For example, in [PaPM86] the optimization is performed for a DFG scheduled and preallocated, and binding is performed to minimize the clock cycle.

The problem of choosing an initial value, and later on optimizing it, is the necessity to obtain a good initial solution to get good final results. Also, the new selection may need a new scheduling, and then a new allocation.

On the other hand, the use of integer divisions may lead to very low cycle times, and the size of the CU can become unacceptably large. Moreover, most of these systems do not take into account the delay due to wiring; only Weng [WePa91] considers it, but only when the floorplanning has been done, in order to modify the design when the time constraints have not been met. This can lead to circuits that do not match users constraints.

8.6.1 Design Model and Algorithm

The design model used by our system is shown in Figure 8.13, where only point to point interconnections are allowed. One level of multiplexers is assumed at the register inputs and another one at the FUs inputs. In this model, we consider multiplexers implemented with two levels of gates, so their delay is approximately constant and independent from the number of inputs. A typical operation involves operands being read from registers, an operation performed on them, and the result stored in a register. The operation delay can go through several control steps. This delay is computed as:

$$delay = t_{oper} + t_{store} + t_{prop} + 2\,t_{mux}$$

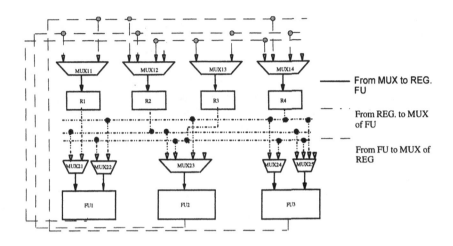

Figure 8.13. Design model and interconnections involved in an operation execution

t_{oper} is the delay of the operator and depends on the FU that will implement it; t_{mux} is the delay of a multiplexer and t_{store} the delay of the register. These three delays depend on the library and they are known. Also, t_{prop} represents the signal propagation, and it depends on four delays: two of them from the multiplexers outputs to the registers (t_{mux_REG}) or FUs (t_{mux_FU}); the other two from the registers (t_{REG_FU}) or FUs (t_{FU_REG}) outputs to FUs or registers respectively, through multiplexers if necessary. These delays are unknown and the methods to estimate them are explained next.

In section 8.5 we obtained that the delay of an interconnection depends on R, C, R_t and C_t. The values for C_t and R_t only depend on the design technology, but R and C also depend on the interconnection length. Moreover signal propagation in the execution of an operation involves four delays, as it is represented in Figure 8.13. Two of these delays are transmissions from multiplexers to FUs or registers (t_{mux_FU} and t_{mux_REG}) that correspond to close interconnections.

The other two kinds of interconnections, t_{REG_FU} and t_{FU_REG} can belong to any of the two types of point to point interconnections, *close* or *distant* in the terminology we have used. We can not know the real interconnection length until placement and routing have taken place, so we need to perform an estimation. A possible solution to make sure the right electrical behaviour of the circuit, is performing an estimation of the **worst case** interconnection length. This maximum length, L_{max}, sets the maximum delay of an interconnection. It is equal to the half perimeter of the circuit, and it depends on the total circuit area by:

$$L_{max} = 2\sqrt{Area}$$

Therefore, the interconnection delay for these two kinds of interconnections is estimated as the delay of an interconnection L_{max}, and then:

But L_{max} depends on the circuit area, that is not known until HA has taken place, and the interconnection delay should be estimated before scheduling in order to select a right clock cycle, so we need to estimate the total circuit area. However, any estimation of the circuit area before scheduling can not give acceptable results because the information of the circuit is incomplete. We propose another solution to solve this problem, that has been included in our HLS system [SMTH95].

$$t_{prop} = 2\,t_{L_{max}} + 2\,t_{L_{close}}$$

A.- Estimating the signal propagation delay tprop.

An initial clock cycle is estimated without considering the interconnection delay, and using the algorithm presented in [MFTS96], that considers the DFG, the module library and the user constraints. Then, a list scheduling is performed, using for implementing each operation the fastest module in the library, and the maximum area for this scheduling is estimated. With this value, we estimate the maximum value of the half perimeter of the circuit $L_{max}List\_Sche$, and the signal propagation delay is calculated with:

$$t_{prop} = 2\,t_{0.9_{L_{maxListSche}}} + 2\,t_{0.9_{L_{close}}}$$

B.- Estimating the clock cycle, scheduling and HA.

In this step the clock cycle is selected using [MFTS96], but now taking into account the interconnection delay estimated before. Then a new scheduling and a HA are performed, and the final area of the circuit, *A_final*, is calculated using the area estimation method explained before, which has an error lower than 5%.

C.- Calculating tprop *and selecting the final clock cycle.*

Using *A_final*, it is possible to calculate the half perimeter of the circuit L_{final}, and to obtain the new signal propagation delay t_{prop_new}:

$$t_{prop\_new} = 2\,t_{0.9_{L_{final}}} + 2\,t_{0.9_{L_{close}}}$$

This value can be used to generate t_{dif}, that is the difference between the value estimated in phase A and the new value

$$t_{dif} = t_{prop} - t_{prop\_new}$$

It is possible to decrease the clock cycle estimated in phase B in a value equal to t_{dif} divided by the maximum number of steps that any FU in the circuit needs to execute an operation. In this way, the scheduling and HA performed before are valid, though the clock cycle changes.

On the other hand, we consider that the control unit works in parallel with the data-path, and so it is only necessary to account for the delay of the control signals. This implies the addition of the delay of one interconnection to the clock cycle. With these assumptions it can be guaranteed the correct time simulation of the circuit.

8.6.2 Example

To show the goodness of this algorithm we present the results for the Fifth Order Elliptical Filter [HLSB91]. The module library is presented in table 8.2. In this example we will assume that area minimization is more important than time minimization. In order to select the clock cycle considering interconnection delay, we follow the phases presented in previous section.

module	op	area	delay
Adder	add	o.o54mm^2	30ns
Adder	add	o.21mm^2	24ns
Multiplier	x	2.3mm^2	138ns

Table 8.2 Module Library

A.- Estimating signal propagation delay t $_{prop}$.

According to [MFTS96], we have to introduce a parameter α, that measures time priority versus area priority. As area minimization is more important than execution time minimization, we choose $\alpha = 0.3$. For this value using the module library presented in table II and the algorithm presented in [16], we obtain a clock cycle of 69ns. After list scheduling and allocation we obtain a total area equal to 8,6 mm2, and $L_{maxList\_Sche} = 5,858$mm, with an associated delay equal to 1,412ns. The delay of a short interconnection for 1μ technology is

$$t_{0.9\ Lclose} = 0,127\text{ns}.$$

Then:

$$t_{prop} = 2 \times 1.412ns + 2 \times 0.127ns = 4ns$$

B.- Estimating the clock cycle, scheduling and HA.

The new clock cycle obtained for this library considering the signal propagation delay is:

$$t_{cycle} = 71\text{ns}.$$

C.- Calculating tprop *and selecting the final clock cycle.*

After HA we obtain $A\_final=4,6\text{mm}^2$, so $L\_final=4,294\text{mm}$. The delay associated to this length interconnection is

$$t_{0.9}(L\_final)=1,051\text{ns},$$

and

$$t_{prop\_new}=2,356\text{ns}.$$

Consequently

$$t_{dif}=4ns-2,356ns=1,644\text{ns}.$$

The maximum number of steps for any FU in the data-path is 2, so it is possible to decrease the clock cycle in

$$\frac{t_{dif}}{2}=0,822ns$$

This value is lower than 1, so the new clock cycle would be:

$$t_{cycle\_new}=t_{cycle}.$$

In order to take into account the control signal delay, we should add 2,356ns (3ns) to the final clock cycle.

8.7 Conclusions

In this chapter, estimations of area and delay of an integrated circuit implemented with standard cells have been studied, in order to help a HLS tool to take the right decisions. Area estimations are focused on interconnection area since the standard cells area is known. A complete model has been presented based on three different classes of interconnections: close, distant and multimodule. Integration of the estimation technique in a HLS tool has been explained and the validity of predictions has been justified. Interconnection delay estimations have also been studied to compute the propagation time and as a consequence, the possibility to select the proper clock cycle.

References

[DeNe89] S. Devadas, R. Newton. *"Algorithms for hardware allocation in data path synthesis"*, IEEE Trans. Computer-Aided Design, vol. 8, pp. 768-781, July 1989.

[ES293] ES2 ECPD10 Library Databook, 1993

[HLSB91] High Level Synthesis Benchmarks, 1991

[Kuo85] B. C. Kuo, *"Automatic Control Systems"* Prentice-Hall International Editions

[MFTS96] H. Mecha, M. Fernández, F. Tirado, J. Septien, D. Mozos, K. Olcoz *"A Method for Area Estimation of VLSI Integrated Circuits in High Level Synthesis"*, IEEE Trans. on CAD of Integrated Circuits and Systems, vol. 15, n°2, pp. 258-265, Feb. 1996.

[MoHF96] R. Moreno, R. Hermida, M. Fernández. *"Register Estimation in Unscheduled Data Flow Graphs"*. ACM TODAES, Vol. 1, no.3, pp. 396-403. July 1996.

[Pang88] B. M. Pangrle, "Splicer: A heuristic approach to connectivity binding", *Proc. 25th Design Automation Conf.*, pp. 536-541, June 1988,

[PaPM86] A.C. Parker, T. Pizzaro and M. Mlinar, "MAHA: A Program for Datapath Synthesis", *in Proc. of the Designs Automation Conference*, pp 461-466, 1986.

[Saku83] T. Sakurai *"Approximation of Wiring Delay in MOSFET LSI"* IEEE Journal of Solid-State Circuits, Vol. SC-18, n° 4, pp. 418-426, Aug. 1983.

[SMTH95] J. Septién, D. Mozos, F. Tirado, R. Hermida, M. Fernández, H. Mecha *"FIDIAS: an Integral Approach to High Level Synthesis"* IEE Proceedings Circuits, Devices and Systems, vol. 142, n°4, pp. 227-235, August 1995.

[WaPa95] C. Wang, K.K.Parhi, *"High Level DSP Synthesis Using Concurrent Transformations, Scheduling and Allocation"* IEEE Transactions on Computer-Aided Design of Integrated Circuits and Systems, vol. 14, n° 3, pp. 274-295, March 1995

[WeEs94] N.H.E. Weste, K. Eshraghian *"Principles of CMOS VLSI Design"* 2nd Edition, cap. 4, 1994

[WePa91] J. Weng, A. Parker, "3D Scheduling: High Level Synthesis with Floorplanning", *28th ACM/IEEE Design Automation Conference*, 1991, pp. 668-673.

9 TEST SYNTHESIS OF DIGITAL SYSTEMS

Pablo Sánchez
Víctor Fernández

9.1 Introduction

Since their appearance, in the late fifties, the complexity of integrated digital systems has increased greatly. Gordon Moore predicted, at the beginning of the sixties, that the number of transistors in a chip would double approximately every two years. This law, known as Moore´s law, has been fulfilled since then [Sap84], in fact, the complexity of integrated circuits is now reaching a billion transistors. The development of such complex circuits is possible thanks to the use of a series of methodologies and programs which aid the engineer in all the phases of the design process.

Traditionally, this process was divided into two steps: the design step and the test step. The objective of the design step is the development of a physical implementation of the digital system, whose specification constitutes its starting point. This process is made up of various steps, as can be seen in figure 9.1. In the first place, a schematic capture is carried out, obtaining a logic netlist. This netlist is simulated with the object of verifying the function of the circuit. Once the logic circuit which verifies the specifications has been obtained, the generation of the layout masks is initiated through edition or placement and routing programs. These layouts are sent to the foundry where the application specific integrated circuits (ASICs) are fabricated. A test set is then needed which allows the determination of whether the chip is defective or not. This is the objective of the test step. Traditionally, this step starts from the logic netlist generated during the design step and its objective is to determine the set of test vectors which will be programmed in the automated test equipment (ATE) that identifies the defective chips.

J.C. López et al. (eds.), Advanced Techniques for Embedded Systems Design and Test, 201-230.
© 1998 *Kluwer Academic Publishers.*

Traditionally, the design and test steps were carried out by different teams of engineers, normally from different departments within the company.

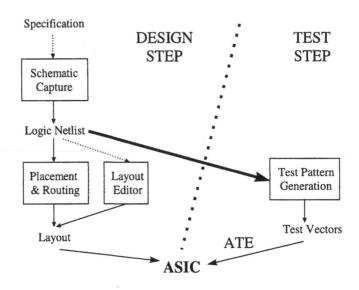

Figure 9.1. Classical Design Methodology

As the complexity of the integrated circuits increased, the complexity of the test step increased much faster than that of the design step. As a consequence, in the mid-eighties, the cost of testing a chip was of the same order as that of designing it. The increasing complexity of the test not only implied an increase in cost of the test step, but also the appearance in the marketplace of badly verified devices. For example, in 1987 Intel had to replace several hundred thousand chips of the 80386 microprocessor that had been incorrectly verified. This error caused an estimated cost of $7 million for Intel. Nor did the test problems stop there: two years later, in 1989, similar problems occurred with the i486 microprocessor. As the managers of Intel pointed out, the problem was not that the microprocessor had been inadequately tested, but was the nature of a complex chip: it was not possible to develop a test program that cover all the possibilities [Ele87].

This type of problems obliged the reconsideration of the role of the test at the design step, giving rise to the appearance of Design-For-Test techniques (DFT). The basic idea was to introduce, at the design step, certain logic structures which facilitate the test step. These logic circuits permitted the reduction in test cost and the increase of the reliability of the defective chip identification process but with the drawback of increasing both the chip area and the critical path. This loss of performance meant that the designers showed some opposition to its general use.

In a parallel way, at the end of the eighties, one of the phenomena which has had most influence on the design process took place: the appearance of commercial automatic synthesis tools. These programs substantially modified the design process, as can be seen in figure 9.2.

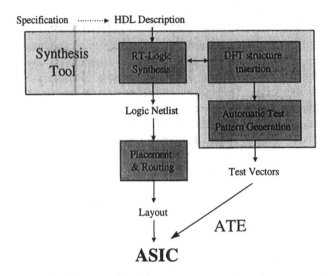

Figure 9.2. Synthesis-based Design Methodology

Starting from the initial specifications, the designer generates a register-transfer level description using hardware description languages like VHDL or VERILOG. This description is synthesized by the RT-logic synthesis program giving the logic netlist of the circuit. At the beginning of the nineties, the commercial synthesis tools started to integrate modules for the automatic insertion of the above-mentioned DFT structures. This integration permitted the resolution of some of the above-mentioned problems. On the one hand, the synthesis tool offers the possibility to synthesize the circuit with DFT structures in order to improve its performance. On the other hand, as they are integrated in the same environment, the design time and the possibility of error in the design stage are reduced greatly.

The first consequence of this integration was the disappearance of the test stage as an independent process, giving rise to a new process: test-synthesis. One of the first definitions of this process was given by B. Bennets in 1994: "Test synthesis is the automated synthesis of hardware testability structures during the process of design, working from Hardware Description Language specifications"[Ben94]. Today, all the commercial synthesis tools offer this type of product, totally integrated into the design flow. So, from the commercial-tool point of view, test synthesis means the practicality of DFT. The evolution of the commercial synthesis tools has been very important. The latest versions permit the initiation of the synthesis at algorithm level as they incorporate the behavioral synthesis tool (Figure 9.3).

These tools have reduced the test cost 10% to of the total cost, in the case of chips, and to 30% in the case of the board. The principal drawback is that besides needing more area and clock cycle time, these techniques increase the synthesis time producing an increase in time-to-market. This parameter is very important because a six-month delay in time-to-market could cut profits by 34%[Lee97].

In order to shorten the development cycle, high-level synthesis (or behavioral synthesis) is a promising technological solution. This methodology can include the testability cost as a synthesis cost. As a consequence, High-Level Tests Synthesis (HLTS) came into being. For a given test strategy to be used in the design, high-level test synthesis is able to explore the synthesis freedom provided at algorithmic level to derive an inherently testable architecture at low or even no overhead (M. Lee Definition [Lee97]), It is sure that future versions of commercial synthesis tools will incorporate these techniques.

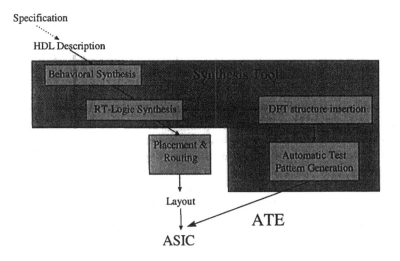

Figure 9.3. Current Design Methodology

The main objective of this chapter is to show how the process of automatic synthesis of digital systems is modified in order to facilitate their testing. In this way, we try to improve the understanding of the process by the future users of synthesis tools and so, to improve the quality of its final design. The chapter is divided into 5 sections:

1. This brief description of the historical evolution of the design process and its relation to testing.

2. An analysis of the principal techniques of DFT used in synthesis tools.

3. Current state of commercial RT-logic synthesis tools for testing.

4. The immediate future of test synthesis: high-level test synthesis.

5. Open problem: system test-synthesis. Test synthesis is an immature topic which is strongly related to the evolution of the automation of the design process. At the moment, the automation of the first step of this process (system synthesis) is under research. The lastest tendencies in this field and their relation to the testing will be presented in this section.

9.2 Test Methodologies

The possible defects which can be produced in the fabrication process are modeled at logic level as faults. The most usually fault model is the stuck-at model. In this model, the faulty circuit net is forced to a fixed '0' or '1' logic value. The objective of this test step is to determine the test vectors which allow us to force a certain logic value in the net under test and propagate the net value to an output. The capability to force a certain logic level in a net of the circuit is denominated controllability and the capability to propagate this value to the outputs of the circuit is called observability. For a circuit to be easily testable, it is necessary that all its nets have high controllability and observability.

The combinational circuit test is not a problem: all single stuck-at faults will be testable if the combinational circuit is not redundant [Abr90]. However, the test of sequential circuits is a very hard problem. Its complexity depends on two factors: the sequential depth (maximum number of flip-flops in an acyclic path from an input to an output) and the sequential loops (cyclic paths through flip-flops). In the example in figure 9.4, a circuit with a sequential depth of 2 and a sequential loop can be observed.

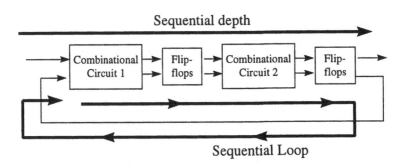

Figure 9.4. Sequential Depth and Sequential Loop

When generating test vectors for a sequential circuit, the complexity is considered to depend linearly on the sequential depth and exponentially on the sequential loops. Figure 9.5 shows the results obtained with an Automatic Test

Pattern Generator tool (ATPG): HITEC [Nie91]. As an example, the description of the Elliptical Wave Filter (EWF) has been used. This description has been used as a high-level synthesis benchmark [Lee97]. With the aim of reducing the complexity of the data-path, the registers have been defined with 3 bits. From the analysis of the description it is concluded that the system has 27 sequential loops and the maximum sequential depth is 9. To verify this system, HITEC generates 95 test vectors, capable of detecting 69.68% of the stuck-at faults (fault coverage). This coverage is too low and would permit a very large number of defective chips to reach the market. In order to reach an adequate fault coverage, it is necessary to facilitate the task of the ATPG by modifying the design, in such a way that the controllability and observability of the circuit nets increase.

* Data-Path width	3
* Maximun sequential depth	9
* Number of sequential loops	27
* ATPG	HITEC
* ATPG execution time	1292 sec.
* Number of test vectors	95
* Fault Coverage	69,68%
* Fault Efficiency	70,72%

Figure 9.5. Sequential ATPG Example

These methodologies of design modification are known as design for test (DFT) techniques [Abr90]. These techniques can be classified in 4 groups:

- **Ad-hoc techniques**. These methodologies try to improve the controllability and observability of the system by using some heuristic solutions which depend on the type of system.

- **Scan techniques**. These methodologies are independent of the type of system. In these techniques, through use of special flip-flops the sequential depth and/or the sequential loops of the circuit are eliminated/reduced, thus obtaining a fault coverage of nearly 100%.

- **Self-test methodologies**. In contrast to other techniques of DFT, these allow the introduction into the system of a set of circuits capable of determining by themselves, whether the circuit is fault free. Furthermore, these structures can be used not only for verifying the device after manufacture but also for maintenance during its normal operation in the system in which it is integrated.

- **Board-Level technique**: Boundary scan. This methodology is a standard extension of the scan techniques to boards and their embedded devices.

The ad-hoc techniques are a set of heuristic rules which allow the improvement of the testability of the circuit. The most important are:

- **Insertion of test points**. The objective is to access the low-controllability nets from the primary inputs of the circuit, and propagate the values of the low-observability nets to the primary outputs of the system. With the aim of not increasing the number of pads of the chip, the pins of the chip are usually multiplexed. Although the principle is an ad-hoc technique (without general application) which depends on the ability of the designer to determine the best set of test points, a series of automatic insertion techniques has been developed which will be presented in a subsequent section.

- **Definition of the initialization**. In sequential circuits it is very important to know the initial state of the system. With this aim, the flip-flops are modified so they can be initialized through a reset signal which forces the system to a known state.

- **Partition** of large counters, shift registers and combinational logic blocks in smaller, more easily testable units.

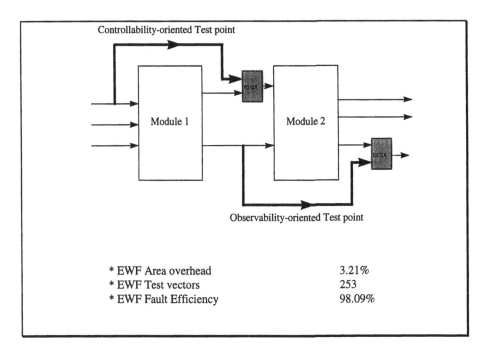

Figure 9.6. Test Point Insertion Example

An example of application of the ad-hoc techniques is shown in figure 9.6. This figure shows the results of introducing test points in the EWF previously shown. As can be seen, the insertion of test points implies the introduction of multiplexors in the circuit while not increasing the number of pins of the chip. This produces an area overhead of 3.2%. Thanks to this small increase of cost, the coverage of faults with the HITEC ATPG reaches over 98%.

The advantages of the ad-hoc techniques are that an at-speed test can be done, that few test vectors are necessary and that the area overhead is small. Nevertheless, the major drawbacks are that they depend strongly on the system type and that the fault coverage which can be obtained is unpredictable a priori. These problems are overcome by scan techniques.

The basic idea of scan techniques is to define an operation mode of the flip-flops which is different to the normal mode: the scan-chain mode during the test. During this mode, the flip-flops of the system are configured as one or various shift registers (scan chains), allowing us to assign any value to a flip-flop (maximum controllability) and to know its value (maximum observability). The most used technique to insert the scan chains in a system is the location of multiplexors at the inputs of the flip-flops as can be observed in Figure 9.7.

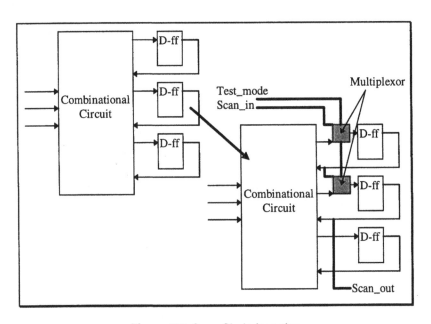

Figure 9.7. Scan Chain Insertion

The technique which inserts all the flip-flops of the system in scan chains is denominated full scan. The principle advantage of this technique is that in test mode the system behaves like a combinational circuit. Coverages of 100% being obtained in absence of redundancies. However, the area overhead introduced by the scan chains can be very great. An alternative technique, partial scan, consists of inserting in the scan chain only a certain number of flip-flops. The most used selection criterion is the breaking of sequential loops. In this way, in test mode, the system behaves like a sequential circuit but without sequential loops which greatly reduces the complexity of the test. Figure 9.8 shows the results of applying these techniques to the description of the EWF.

Scan Type	Full scan	Partial Scan
Scan Chain	36 flip-flops	24 flip-flops
Selection Criteria	-	Break sequential loops
Area Overhead	11%	7.7%
Test vectors	2589	8229
ATPG Time (sec.)	12.6	1536
Fault Coverage	100%	100%

Figure 9.8. EWF Scan Insertion Results

As can be see, coverages of 100% are obtained, with a greater overhead than with ad-hoc techniques, and a very large number of test vectors. The difficulty of testing memories, and the reduction of speed of operation of the system during the test mode, which means that time-dependent faults cannot be detected, are other disadvantages which should be taken into account.

The Built-In Self Test (BIST) techniques are based on the fact that a high percentage of faults can be detected simply by applying a sufficiently large sequence of random values at the inputs of the circuit. This methodology implies the redesign of the system's registers so that in a test mode they can behave as a Random Test Pattern Generator (RTPG) or a Multiple Input Signature Register (MISR). As can be seen in Figure 9.9, the basic idea is that in test mode the RTPG generates a sequence of pseudo-random values and applies them to the circuit to be verified (in the example, a static RAM of 1Kx16). The responses of the circuit under test are compressed by the MISR register, generating a signature as an output (signal Bist-result in the example) which indicates whether the circuit is fault free or not. As can be observed in the example, the functional test of the memory requires the application of 9260 test vectors to the device and the checking of 4096 values. Using BIST techniques, it is only necessary to change one bit (Bist_test) and wait 8200 clock cycles, after which the output Bist_result will show the result of the test for 13 clock cycles. The increase in area of this technique is small (2.54% in the example).

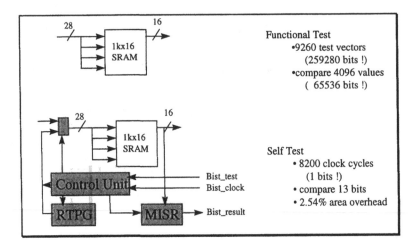

Figure 9.9. BIST Example

Besides these advantages, the BIST techniques carry out at-speed test and also a test on-field. The main disadvantages are the increase of the critical path and the power dissipation as well as the relatively low fault coverage. This low coverage is due to aliasing phenomena, to the presence of faults which are resistant to this type of technique and to the limitations of the size of the pseudo-random sequences which are made for practical reasons.

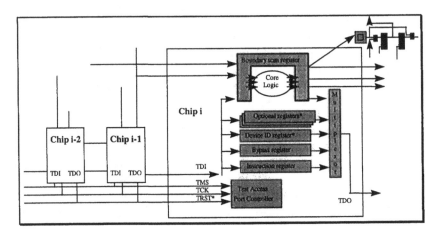

Figure 9.10. Boundary Scan Example

The IEEE 1149.1 Boundary Scan Standard [Std90] came into being with the aim of facilitating the test of the boards and the components soldered to them. The standard defines a series of procedures which permit the checking of the connections of the board and also the access to the pins of the soldered components and even includes access procedures to internal registers or to DFT structures. Figure 9.10

shows the structure of a board with boundary scan. All the components form a part of the scan chain. Each component is designed in accordance with the standard. In this way, it is possible to access the pins of the chip and the internal registers that are necessary (e.g. the full scan registers) from the exterior of the board connections.

9.3 RT-Logic Test Synthesis

The principal objective of this section is to show the integration of the test step in the commercial RT-Logic synthesis tools. These programs normally use various DFT techniques, presented in the previous section: scan (full and partial), boundary scan and BIST techniques. The reason for this choice is that these techniques guarantee a high coverage independently of the system, as well as being relatively easy to introduce in the synthesis flow of the tool. Depending on the way the DFT structure are introduced, the commercial synthesis tools can be classified in three types.

The first type of tools inserts only scan-type structures (full scan and partial scan) and boundary scan, once the process of RT-Logic synthesis is over. As can be seen in figure 9.11, the integration of the DFT structures obliges a re-synthesis of the system, with the consequential increase in design time and loss of performance. This scheme has been adopted in CAD frameworks like Cadence.

A second type of tools permits a pre-insertion of the DFT structures, before carrying out the synthesis, as can be seen in Figure 9.12.

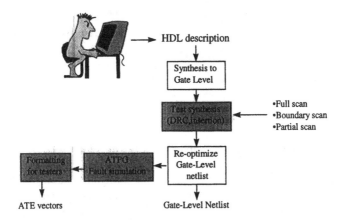

Figure 9.11. Integration into a Design Flow: Type 1

In this way, the necessity of re-synthesizing the circuit after the insertion of DFT structures can be avoided. An additional possibility offered by this type of program is the capacity to carry out hierarchical insertions, that is, to insert scan-type structures only in a particular block or system model; this is of great interest when designing reusable modules. The type of DFT structures which are supported by these programs are the same as those of type 1. The Test Compiler program of Synopsys is an example of this type of environment.

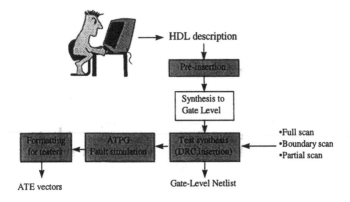

Figure 9.12. Integration into a Design Flow: Type 2

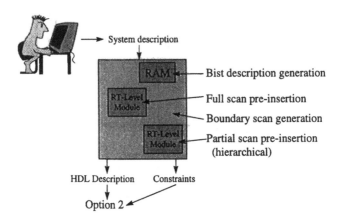

Figure 9.13. Integration into a Design Flow: Type 3

The third type of tools offers the possibility of mixing different DFT methodologies in the same chip. Besides the insertion, the synthesis tool creates the protocols and procedures that permit the verification of the system with multiple test strategies. Figure 9.13 shows an example of how this type of tool works. In the example, the user has decided to verify the internal RAM memory using BIST techniques, one module with full scan and the other with partial scan, and to manage all the DFT structures through boundary scan. The synthesis tool does not only insert the requested structures (normally using methodologies similar to the type 2 tools), but also generates the protocol which allows the chip to be tested on board, through boundary scan. An example of this type of tool is the Logic Vision environment.

The insertion of DFT structures is carried out in approximately the same way in all tools. Firstly, a generic DFT rule checker (DRC) identifies structures which might cause problems with the selected test strategy like asynchronous feedback, internally generated clocks, etc. Then, the flip-flops are modified to insert the scan chains, or the synthesizable descriptions of the BIST structures and/or of the modules forming the implementations of the standard boundary scan are introduced. In general, the tools permit the selection of the number of scan chains, the flip-flops making up the chain and their order. Details of the IEEE 1149.1 standard [Std90] such as specific instructions or internal registers included in the boundary scan are also allowed. In this last case (insertion of internal registers in boundary scan), clock gating is added, which must be taken into account when analyzing the performance of the final circuit. Once the DFT structures are inserted and the circuit is synthesized, the tool itself takes charge of generating the test vectors and the protocols necessary for application to the chip which is being designed.

During the RT-logic synthesis the commercial synthesis algorithm that most directly affects the testability of the circuit is the logic synthesis of the combinational part. This is initiated once the architecture is defined and the flip-flops of the system are implemented. During this synthesis procedure, the tool introduces redundant gates with the aim of reducing the critical path, the area or the interconnections. These redundancies affect the system testability negatively, as was mentioned in the introduction. For this reason, as well as area restrictions, working frequency and power consumed, the user must specify the minimum coverage which the system's combinational logic blocks must have. This value will be used to limit the number of redundant gates and so improve the testability. In general, the commercial synthesis tools fix the fault coverage in the combinational logic between 95% and 98%. These values ensure that the impact of the redundant gates on the complete system testability is low.

The effect of the other RT-logic synthesis techniques, like re-timing, on the testability of the system is an area for study. The use of other DFT techniques like automatic insertion of test points does not seem to be a short-term objective of the commercial environments even though these include analysis programs capable of determining the zones of low testability of the circuit and also capable of inserting in-board test points (like the case of the Mentor environment).

9.4 High-Level Test Synthesis

In contrast with the last section, in which the impact of testability on the synthesis algorithms was relatively low, in high-level synthesis the consideration of this parameter during the exploration of the design space by the synthesis tool leads to important improvements during the test of the system.

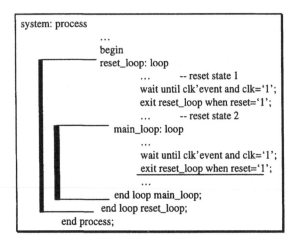

```
system: process
              ...
            begin
            reset_loop: loop
                ...         -- reset state 1
                wait until clk'event and clk='1';
                exit reset_loop when reset='1';
                ...         -- reset state 2
            main_loop: loop
                ...
                wait until clk'event and clk='1';
                exit reset_loop when reset='1';
                ...
            end loop main_loop;
            end loop reset_loop;
        end process;
```

Figure 9.14. Algorithmic Description Style

The consideration of testability starts in the description of the system at algorithmic level including the explicit definition of the reset signal in the descriptive style, thus allowing the system to be brought to a known state, as can be observed in Figure 9.14. In this way, each time an update of the system's signals (wait statement) is specified, a check of the reset signal is included which brings the system to the reset loop (external loop). This descriptive style is used by the Synopsys Behavioral Compiler.

The area of High-Level Test Synthesis (HLTS) is still at the investigation stage, there being no commercial tools dealing with it. The investigation lines in HLTS can be divided into two big groups depending on the test strategy which they follow after the synthesis process: HLTS oriented to self-test and HLTS oriented to the application of external test. The latter, normally applies the test in an ATE with test vectors previously calculated with an ATPG. In some cases, they are supported by a DFT technique.

The result of the high-level synthesis is an RT-level structure made up of two well-differentiated blocks: a controller and a data path (see Figure 9.15). A very common supposition in the majority of the HLTS tools included in the two tendencies is to suppose that the controller and the data path can be tested independently. To do this, it is necessary to multiplex the signals which interconnect both blocks towards primary inputs and outputs. In most cases, this can have a high cost. The supposition of independent testing for controller and data path has meant that most of these tools work by focusing only on the data path. Recent approaches have demonstrated the strong influence of the controller on the testability of the complete circuit. We will now focus on the two above-mentioned tendencies in more detail.

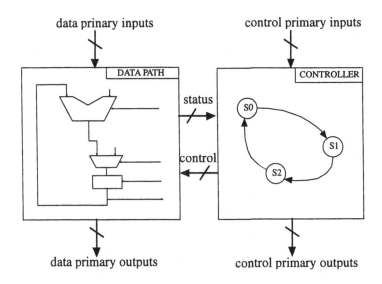

Figure 9.15. RT-level Structure

9.4.1 High Level Test Synthesis For BIST

We will now take a look at the three most representative systems of high-level synthesis for BIST which have been developed in recent years. Before doing so, we will explain the common problem which they face.

The BIST strategy used by all the systems we will study is based on the use of BILBOs. A BILBO is a structure which can be configured to work as a normal register, as a pattern generator, as a signature analyzer or as a shift register. During the design of the data path of an RT architecture, if we want to introduce a BIST structure, as a first approximation we can consider substituting the registers of the data path for BILBOs. However, this is not always possible. The presence of self-loops in the data path impedes our aim. To illustrate this we shall look at Figure 9.16. As can be seen, the BILBO register provides the operational unit with data and at the same time the output of this unit is stored in the register itself. This is what is called a self-loop. With this configuration, in order to inject patterns into the functional unit we must configure the BILBO as a pattern generator. However, at the same time, in order to store the response of this operational unit we must configure the BILBO as a signature analyzer. The systems we will review in this chapter have the objective that the synthesis result does not have this type of structures.

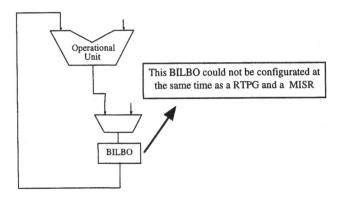

Figure 9.16. Problem caused for the BIST by the presence of a self-loop

The three systems we shall see are: SYNTEST [Hrm92], RALLOC [Avr91] and SYNCBIST [Hrr94].

9.4.1.1 SYNTEST

SYNTEST [Hrm92] is a high-level synthesis tools for self-test, developed in Case Western Reserve University. Starting from the SDFG (Scheduled Data Flow Graph) the final result is a data path with BIST structures. The technique applied is an allocation algorithm with the following three steps:

- **Step 1.** A Testable Functional Block (TFB) is defined to be that formed by a functional unit and a register at its output. The first step of allocation is the assignment of each operation and its output variable of SDFG to a TFB. The only restriction in this first step is that self-loops are not produced. In this way, it can be ensured that the data path generated will be testable through BILBOs.

- **Step 2.** This is a refinement of step 1. The objective is to minimize area and delay of the data path generated, through the sharing of resources. These resources will be shared when there are no incompatibilities and when no self-loops are generated. Certain cost measures are used in order to guide the refinement stage. The result of this stage is a data path to which BIST can be applied through BILBOs and with a sequential depth of 1 (between pattern generator and the signature analyzer there is only combinational logic).

- **Step 3.** This is an even greater refinement than in step 2 based on the functional testability of the modules [Chu91]. An analysis of the functional testability of each module can lead to the elimination of certain unnecessary pattern generators or signature analyzers. In this way, some registers will not convert to BILBOs. The BIST scheme generated now has a sequential depth greater than 1.

The published results of this tool show an extra area of about 25% for the BIST of sequential depth 1, and area around 4% for the BIST with a sequential depth of more than 1 with similar fault coverage in both cases. The authors of SYNTEST have even published improvements in the results through RT-level refinements [Hrm93].

9.4.1.2 RALLOC

RALLOC [Avr91] is a register allocation tool whose objective is to include self-test in the data path of the RT-structure generated. It was developed in Stanford University. The main objective is to minimize the number of self-loops during the allocation stage. The algorithm starts from a SDFG; the allocation of operational units has also been carried out. Next, the algorithm extracts an incompatibility graph of the variables, that is, the nodes are variables and these are joined by an arc when they cannot share register. Starting from the graph, the algorithm applies a graph coloring heuristic in an attempt to minimize the number of self-loops. The registers which remain within a self-loop will be substituted by a CBILBO and those outside are substituted by BILBOs. A CBILBO is an element that can act as a pattern generator and as a signature analyzer at the same time. Its use is restricted as much as possible since its cost in area is significantly greater than a BILBO. The cost of the final RT architecture is estimated with the following cost measure:

$$Cost= 20*(NSAreg) + 35*(SAreg) + mux\_in + int\_con + ctl\_sig$$

Where NSAreg is the number of BILBOs, SAreg is the number of CBILBOs, mux_in is the number of inputs to multiplexors, int_con is the number of interconnections and ctl_sig is the number of control signals. The experimental results show that the circuits synthesized with RALLOC have a lower cost.

9.4.1.3 SYNCBIST

SYNCBIST [Hrr94] is a high-level synthesis tool developed in the University of San Diego. Its aim is to synthesize data paths with BIST and to provide a highly concurrent test plan in order to reduce the test time. The final BIST can have a sequential depth greater than 1. Self-loops are avoided with two test sessions (instead of with a CBILBO).

A test path is defined as a piece of circuit through which test vectors are propagated. When two test paths share hardware, they are said to be in conflict. The principal consequence is that they cannot be tested concurrently and this leads to an increase in test time. SYNCBIST attempts to avoid these conflicts during the synthesis process. The main conflict occurs with the presence of a self-loop.

The authors propose some metrics of testability for any intermediary state of the design during the synthesis process. These metrics have to estimate the conflicts which will appear in the final design. The synthesis is guided by the above metrics towards a design which avoids conflicts in the greatest possible number of cases. The experimental results demonstrate that the tool achieves structures, in the majority of that cases, which only require one test session.

9.4.2 High-Level Test Synthesis For ATE

The high-level synthesis systems which we will see in this section have a common aim which is to facilitate the work of the ATPG after the synthesis process. The strategy which they follow is different. In some cases the synthesis is done with the idea of inserting a partial scan in the final design. In other cases no DFT structure is inserted. In another case, the high-level synthesis will facilitate the hierarchical test. Finally, we shall see an approximation in which the controller will be modified with the object of improving the testability of the complete circuit. We shall see the above-mentioned systems separately.

9.4.2.1 PHITS

PHITS [Lee97][Lee93][Le92a][Le92b] is fundamentally an allocation tool. It has been developed in the University of Princeton and it applies test criteria based on the elimination of loops and reduction of the sequential depth in order to improve the yield of the ATPG. Although they also speak about test criteria during the scheduling [Le92b], this part in reality is the least important within the tool. The most important part is the allocation [Lee93][Le92a], and especially the allocation of registers. Then the input to the algorithm is a Scheduled Data Flow Graph (SDFG).

After the allocation of the data path, its RT-structure can be represented through a graph denominated Data Path Circuit Graph (DPCG) very similar to the S-graph of [Che90] at the logic level. In the DPCG each node represents a register and each directed arc represents an operational unit connecting two registers. A sequential path is defined as the directed path between two nodes of a DPCG. The length of this path will be the number of arcs it has. A non trivial cyclic sequential path is defined as that whose length is greater than 1; furthermore none of the registers of the path can be input/output registers. The path is called a trivial cyclic sequential path when its length is 1 (usually called self-loop). The sequential depth between an input register and an output one is defined as the length of the shortest sequential path between them.

In [Che90] it is demonstrated that the sequential depth and above all the cyclic paths (especially the non-trivial ones) affected the yield of the ATPG negatively. Therefore, PHITS carries out the allocation attempting to avoid non-trivial cyclic paths and at the same time reducing the sequential depth.

Two types of loops are considered by PHITS. On the one hand, those which are obviously formed because there are loops at the Data Flow Graph. On the other hand, those formed during the allocation stage on assigning two variables of a path of the SDFG to the same register. This case can be seen in Figure 9.17. Nevertheless, as we shall see later, there are other types of loops that are not being considered.

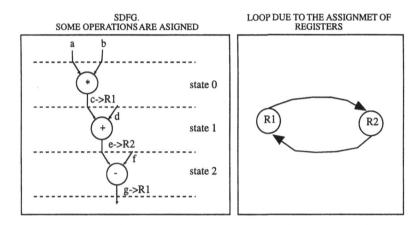

Figure 9.17. Allocation dependent loops

PHITS admits as an input the maximum number of scan registers which can be in the final circuit. When this number is fixed at 0, the final circuit will not have any test structure. In this case the algorithm is denominated PHITS-NS. When the inclusion of scan registers is allowed, the algorithm is called PHITS-PS. PHITS applies heuristics with the aim of obtaining a data path in which the maximum number of loops is reduced as much as possible. At the same time, it attempts to reduce the sequential depth of the resulting RT-structure.

With PHITS-NS the experimental results show that the ATPG yield improves with the architecture resulting from this algorithm compared to the yield with the circuits resulting from other synthesis tools which do not take into account the test criteria. The results of ATPG with the circuit resulting from PHITS-PS are logically better than those of PHITS-NS. The results also show how, if partial scan techniques are applied to the circuits resulting from other synthesis systems, this will lead to a circuit with more scan registers than the number selected by PHITS-PS. Additionally, the ATPG results will be poorer than with PHITS-PS. It is important to highlight that the results presented and the tests carried out with PHITS include the data path alone. That is, the controller has not been included. Therefore, the load signals of the registers and those of selection of the multiplexors which will in reality come from the controller are included as primary inputs.

9.4.2.2 BETS

BETS [Dey93] is the high-level synthesis system for partial scan developed in NEC C&C Research Laboratories. The BETS system is applied to the register file model. It is the RT architecture used by Cathedral [Man86] and by HYPER [Rab91]. In this structure, the registers are grouped into files of registers which only give data to a single operational unit. However, the files of registers can receive data from any operational unit. In Figure 9.18 we can see this type of topology.

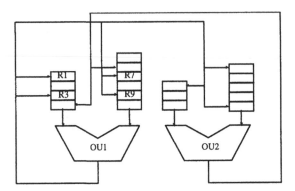

Figure 9.18. Section of data path following the register file topology

For the type of topology considered, the authors distinguish 4 types of loops that can be formed during the synthesis process:

- **CDFG loops**. These are loops formed by those loops which are already present if the CDFG is cyclic.

- **Assignment loops**. These are produced when two operations of the same CDFG path are assigned to the same operational unit. When these operations are consecutive they form a self-loop which will be ignored.

- **False path loops**. These are loops which are not produced in the normal function of the circuit. However, the test vectors will be produced assuming total controllability in the control signals, that is, the same as occurred with PHITS, the control inputs will be considered as primary inputs without testing the controller. In this case the data path can acquire a flow with new loops which are false path loops.

- **Register files cliques**. These are multiple self-loops which are formed around an operational unit. For example, in figure 9.18, the registers R1, R3, R7 and R9 form self-loops with the operational unit OU1. To break the loops formed among them it will be necessary to select for scan all the registers except one.

The starting point is the CDFG. The first step is to break all the loops of this graph with a minimum number of registers. Then, the scheduling and allocation is done attempting to avoid the formation of the above-mentioned loops (not self-loops) using, as much as possible, the scan registers already assigned without increasing the area too much.

The experimental results were obtained using the ATPG HITEC [Nie91] and the partial scan insertion tool at logic level OPUS [Chk91]. The results of ATPG, obtained following the test algorithms described, were compared with the those obtained when carrying out a conventional synthesis trying to achieve maximum speed and minimum area. The results showed that with test criteria the ATPG results were noticeably better with a very low hardware cost. Furthermore OPUS is applied to the resulting circuit without test criteria. When OPUS selected scan flip-flops so that no loops were left (not self-loops) similar ATPG results were obtained to those obtained with test criteria but more scan flip-flops were needed. If OPUS was limited to using the same number of scan flip-flops as BETS, worse ATPG results were obtained.

9.4.2.3 LOOPALLOC

LOOPALLOC is an allocation tool developed in the University of Cantabria [Fe95b][Fe95c][Fe95a][Fer96]. The result of the synthesis tool is a loop-free circuit at logic level applying partial scan with minimum extra area cost.

On the contrary to the two tools presented beforehand, LOOPALLOC uses the complete circuit, including the controller, when performing the analysis and obtaining results. Figure 9.19 shows the different loops that can be formed in a RT structure.

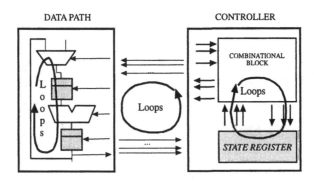

Figure 9.19. Loops in a RT structure

It can be seen that as well as the loops formed through the data path registers, loops are also formed among these registers and the state register of the controller.

On the other hand, there is a self-loop through the state register of the controller whose importance will be seen later on.

A distinction is made between RT loops, like those formed through registers at the RT level, and logic loops like those formed through flip-flops at logic level. LOOPALLOC produces circuits that do not have loops (not considering self-loops) at the logic level. This ensures a high fault coverage. For this, it will select during the high-level synthesis, the registers which break all the RT loops and those RT-self-loops which can form logic loops after the process of RT-logic synthesis. The following classification of RT-loops in the data path is made:

- **Functional loops.** There is a functional loop when there is a dependency between the future value of a register and its current value, or in other words, its input depends on its output. The functional loops are created when two variables of the same path of the DFG are assigned to the same register. On the other hand, a loop will inevitably be formed when there is a cyclic path in the DFG. These last loops are called inevitable loops. The functional loops were the only ones considered by PHITS.

- **Topological loops.** For this type of loops, there is a path in the topology of the circuit between the output and the input of a register. However, in normal function of the circuit there is no dependency between the future value and the current value of the register. But, given that the circuit in test mode will work differently to how it works normally, and/or if we also take into account faults in the control signals, this type of loops can become functional. This last case can be seen in figure 9.20.

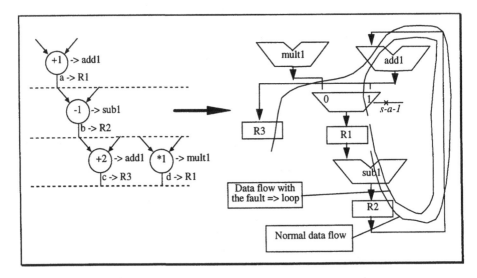

Figure 9.20. Single assignment topological loop

We can distinguish two types of topological loops depending on their origin during the synthesis:

- **Single assignment topological loops.** This type of loops are formed when two operations or two data transfers of one path of the DFG are assigned to the same operational unit or interconnection (bus or multiplexor) respectively. For example, in figure 9.20, a topological loop is formed on assigning the operation *+1* and *+2* to the same adder *add1*. If we consider a stuck-at-1 fault in the control signal of the multiplexor, the loop is closed.

- **Multiple assignment topological loops.** An assignment, in which at least two hardware resources and two paths of the DFG are involved, closes a loop. The most basic case (two paths and two resources) can be seen in figure 9.21 (AE stands for Algorithmic Element and HR stands for Hardware Resource); the extension to more complex cases is easy to see.

The RT loops, that is, the loops which can be formed in the data path, will form logic-level loops after the RT-logic synthesis when the registers are converted into flip-flops and the operational units into logic gates. Given that our objective is to eliminate the logic-level loops, the first premise will be to eliminate the above-mentioned loops.

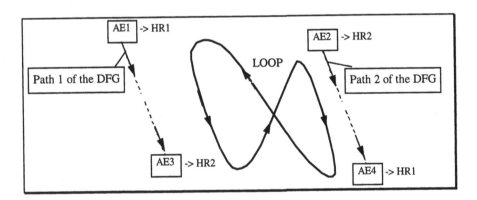

Figure 9.21. Multiple assignment topological loops.

The elimination of the RT loops does not eliminate all the loops that can be formed at logic level. We must also consider the loops which are formed between the data path and the controller (which also form loops on passing to logic level). There is also a special type of RT-level self-loops which form logic-level loops. There are two cases in which RT self-loops form logic level loops. The first of the cases is the self-loop through the state register of the controller. After the RT-logic synthesis, loops can be formed through the flip-flops in which the state register decomposes. LOOPALLOC selects the state register for scan. In this way, the logic loops are broken and furthermore, the loops between the state register and the data path register are broken too (figure 9.19). The second case of logic loops from RT-level self-loop is that of an self-loop through a divider.

Bearing in mind the analysis of loops carried out, and after the selection of the state register for scan, LOOPALLOC carries out the synthesis minimizing area and breaking all the RT-loops previously explained (including self-loops through dividers).

The results obtained with LOOPALLOC were tested with the Synopsys Test Compiler Plus and with HITEC. In both cases, very high fault coverages were obtained with an area less than or equal to the final circuits provided by most of the synthesis tools (including PHITS-PS). Furthermore, the structures synthesized by other tools always needed more scan flip-flops to break all the loops and to be able to obtain similar fault coverages.

9.4.2.4 GENESIS

GENESIS [Bh94a][Bh94b] is a high-level synthesis tool for hierarchical test developed in Princeton University. The strategy it follows consists of using high-level information to facilitate the tasks of generation of test vectors, that is, to justify and propagate the logic values of the internal nodes of the circuit.

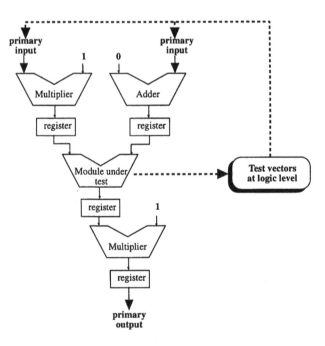

Figure 9.22. The test vectors calculated at logic level for the module to be tested are applied at RT-level taking advantage of the known functionality at this level

For each operational unit of the data path, the test vectors necessary for its logic level test are calculated. GENESIS takes advantage of the functional information which there is at RT-level to get the previously calculated vectors to each operational unit. We will see an example of this in figure 9.22.

If for the structure of Figure 9.22, at logic level an ATPG wants to justify vectors up to the operational unit to be tested, it will do so through the previous operational units but without knowing which are adders and which are multipliers. However, at RT-level we know the function of these modules and we can justify the desired vectors directly from the inputs using the neutral values (0 for the adder, 1 for the multiplier) as can be seen in the Figure 9.22. The same can be applied to propagate the output of the module to be tested.

GENESIS uses the technique described at RT-level to generate the test of the operational units. Furthermore, previously, during the high-level synthesis it is ensured that the RT architecture generated has the justification and propagation paths necessary so that each operational unit can receive vectors directly from the primary inputs and propagate their response up to the primary outputs. When this is not achieved for an operational unit, it adds multiplexors to increase the controllability/observability.

GENESIS includes the controller in the final test. In Figure 9.23, we can see the topology used at RT-level. With this configuration, the test of controller and the data path are carried out separately. However, when the test of the data path is done, the control signals come directly from the controller, that is, the data path is tested following the flow described in the SDFG. A reset input is presupposed in the controller which brings it to its initial state whenever this is desired. To test the controller, with the test input T the state register can receive data from the scan input and take out the control signals multiplexed with the primary outputs.

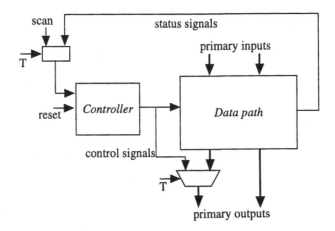

Figure 9.23. RT structure created by GENESIS

The experimental results demonstrate that GENESIS synthesizes the benchmark circuits with test environments for all the operational units. For this, it uses very little CPU time independently of the bit width of the data path generated. This is because the test paths generated do not depend on the bit width (see Figure 9.22)

9.4.2.5 High-Level Test Synthesis based on Controller Redefinition

This work was published by the University of Cantabria [Fer97]. In this work, the great influence of the controller on the testability of the resulting RT circuits is shown. For this, a synthesis is carried out in which the data path is synthesized according to a classic algorithm, without taking into account the test criteria. The controller is synthesized with the objective of favoring the ATPG tasks. To do this, the controller is constructed with different operation modes. With these modes, paths are created from the primary inputs to any operational unit inputs, and from the operational unit outputs to the primary outputs. The modes defined are as follows:

- *Mode 0:* **Normal running mode**. Control signals make the data-path run as described in the high-level description.

- *Mode 1:* **Justification mode**. For any state, multiplexor control signals are fixed so that there is a path from the inputs of any operational unit to the closest primary input. These paths are non cyclic whenever possible. The control signals for the registers are fixed for permanent loading.

- *Mode 2 to n:* **Propagation modes**. For any state, multiplexor control signals are fixed so that there is a path from the output of any operational unit to the closest primary output. The output path of an operational unit can close the output path to another and therefore, more than one mode could be necessary. For the benchmarks tested, the maximum number of propagation modes was two. This means a maximum of four modes and therefore, two test input signals. The control signals for the registers are fixed for permanent loading.

A simple example can be observed in Figure 9.24. In justification mode sm2=1, sm4=1, sm3=0 and sm5=0 to justify the inputs of the operational unit U1 from inport1 and inport2. In order to propagate the output of this operational unit, sm1=1; nevertheless, another propagation mode is needed to propagate the value of U0 (with sm1=0).

We shall see the results obtained for the Facet Benchmark case in Figure 9.25. The first row of the table of Figure 9.25 shows results with the data path alone. In this case, control and status buses are primary inputs and outputs respectively. As can be seen, the results are quite good. The controller is included in the second row of the table. In this case, results do not reach a minimum standard. Therefore, the great influence of the controller on the testability of the whole circuit is clear. Test generation results with the new controller (modif. in the table of Figure 9.25) are shown in the fourth row of the table. In order to have other test generation results for comparison, results for the approach presented in [Dey94] are shown in the third row of the table. In that approach, test points (t. p. in the table of Figure 9.25) are added to the data path and the state register is parallely loaded (p. l. in the table) from primary inputs.

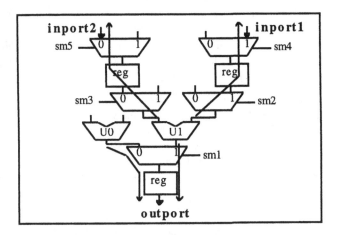

Figure 9.24. Example of test modes

	ATPG time	#vec.	eff. (%)	fc. (%)	extra area (%)	ex. pins
Data Path	203	905	99.94	99.94	-	-
Data Path + Controller	50767	112	35.99	33.28	-	-
Data Path (t. p.) + Control (p. l.)	10870	531	96.15	96.15	3.65	1
Data Path + Control (modif.)	612	1550	99.66	99.51	2.55	2

Figure 9.25. Results for the Facet Benchmark

9.5 Future Trends

The rapid development of integration technologies permits the introduction of more and more complicated systems in a simple chip. The current tendencies imply that future designs will not only have a digital part but that they will include important analog modules and also that part of their function will be carried out by the software executed in the microprocessors embedded in the system. Such heterogeneous systems will be designed making more and more use of embedded cores or pre-designed modules of high complexity like microprocessors, DSPs, etc.

The test of these systems will oblige the use of multiple test strategies, not only of the digital part but also of the analog part and the software. These systems will be designed using co-design techniques like those commented in the previous chapters of this book. From the point of view of test synthesis, two problems merit special attention: the testability oriented co-design or co-test and the embedded core test. Considering the embedded core test, the problem is due to the restrictions imposed by the protection of intellectual copyright of the creator of the module.

Normally, the cores are developed by a company which is not the user. In fact, in recent years, a large number of companies have appeared whose principal business is to sell implementations of the most-used modules like interfaces (ethernet, SCSI, etc), microprocessors (MC6508,Sparc,etc), signal processing circuits (Viterbi decoder, Reed Solomon,...), etc. These companies normally provide encrypted modules which permit the synthesis and simulation of the acquired module, but which do not have sufficient detail to be able to carry out their testing. Therefore, in addition to the encrypted models, the suppliers provide a set of test vectors to be applied to the core along with the expected fault coverage of these vectors.

The problem occurs when carrying out the test of the complete system, in which it is necessary to integrate the test cores of different manufacturers. The solution to this problem is a consensus among the companies which develop cores and the CAD salesmen in the form of a standard, in which the businesses of the sector are now working. The first impressions indicate that the new standard will be similar to the Boundary Scan Standard in terms of philosophy and methodology. It is hoped that this standard will appear in the next years, thus generating the development of new test synthesis techniques which will make use of the standard to verify the complete system. This standard will also affect the development of co-test techniques since the cost of the system test and the expected coverage (principal measures of testability during the system-level synthesis) will be strongly influenced by it.

A new problem is the appearance of part of the function of the system implemented in the form of software that is executed in embedded microprocessors. These modules can also be used to verify other modules of the system. For this reason, during the co-design, the cost of introducing software test programs instead of DFT structures should be evaluated. During the partition stage, the test strategy for each module will be decided on (software or hardware). These techniques are still in the initial investigation stage, their development being very influenced by the evolution of co-design techniques.

References

[Abr90] M. Abranovici, M. Breuer, A. Friedman. *"Digital Systems Testing and Testable Design"*. Computer Science Press. 1990.

[Avr91] L. Avra. "Allocation and assignment in high-level synthesis for self-testable data paths". *Proc. International Test Conference*, pp. 463-472. Aug. 1991.

[Ben94] B. Bennets. "Test Synthesis: the practically of DFT". *Tutorial of The European Design & Test Conference*. March 1994.

[Bh94a] S. Bhatia and N. K. Jha. "Genesis: A behavioral synthesis for hierarchical testability". *Proc. European Design & Test Conference*, pp. 272-276. Feb. 1994.

[Bh94b] S. Bhatia and N. K. Jha. "Behavioral synthesis for hierarchical testability of controller/data path circuits with conditional branches". *Proc. International Conference on Computer Design*, pp.91-96. Oct. 1994.

[Che90] K.-T. Cheng and V. D. Agrawal. *"A partial scan method for sequential circuits with feedback"*. IEEE Transactions on Computers, pp. 544-548, Apr. 1990.

[Chk91]. V. Chikermane and J. H. Patel. "A fault oriented partial scan design approach". *Proc. International Conference on Computer Aided Design*, pp. 400-403, Nov. 1991.

[Chu91] S. Chiu and C. A. Papachristou. "A built-in self-testing approach for minimizing hardware overhead". *Proc. International Conference on Computer Design*, pp. 282-285, Oct. 1991.

[Dey93] S. Dey, M. Potkonjak and R. K. Roy. "Exploiting hardware sharing in high-level synthesis for partial scan optimization". *Proc. International Conference on CAD*, Nov. 1993.

[Dey94] S. Dey and M. Potkonjak. "Non-scan design-for-testability of RT-level data paths" *Proc. International Conference on Computer Aided Design*, pp. 640-645, Nov. 1994.

[Ele87] "Electronics Times". No 404. 16 April 1987.

[Fe95a] V. Fernández y P. Sánchez. "Síntesis de alto nivel para scan parcial". *Actas del X Congreso de Diseño de Circuitos Integrados y Sistemas*, pp. 306-311, Nov. 1995

[Fe95b] V. Fernández, P. Sánchez and E. Villar. "Partial scan high-level synthesis strategy". Second International" *Test Synthesis Workshop*. Santa Barbara, USA, May 1995.

[Fe95c] V. Fernández, P. Sánchez and E. Villar. "A novel high-level allocation technique por test" *Proc. Fourth European Atlantic Test Workshop*. July, 1995.

[Fer96] V. Fernández and P. Sánchez. "Partial scan high-level synthesis". *Proc. European Design & Test Conference*, pp.481-485, March 1996.

[Fer97] V. Fernández and P. Sánchez. *"High-Level Test Synthesis based on controller redefinition"*. Electronics Letters. Vol 33, No. 19, pp. 1596-1597, September 1997.

[Hrm92] H. Harmanani, C. A. Papachristou, S. Chiu and M. Nourani. "SYNTEST: An environment for system-level design for test". *Proc. European Design Automation Conference*, pp. 402-407, 1992.

[Hrm93] H. Harmanani and C. A. Papachristou. "An improved method for RTL synthesis with testability tradeoffs". *Proc. International Conference on Computer-Aided Design*, pp. 30-35, Nov. 1993.

[Hrr94] I. G. Harris and A. Orailoglu. "SYNCBIST: synthesis for concurrent built-in self-testability". *Proc. International Conference Computer Design*, pp. 101-104, Oct. 1994.

[Le92a] T.-C. Lee, W. H. Wolf, N. K. Jha and J. M. Acken. "Behavioral synthesis for easy testability in data path allocation". *Proc. International Conference on Computer Design*, pp. 29-32. 1992.

[Le92b] T.-C. Lee, W. H. Wolf and N. K. Jha. "Behavioral synthesis for easy testability in data path scheduling". *Proc. International Conference on Computer Aided Design*, pp. 616-619. 1992.

[Lee93] T.-C. Lee, N. K. Jha and W. H. Wolf. "Behavioral synthesis of highly testable data paths under the non-scan and partial scan environments". *Proc. IEEE Design Automation Conference*, pp. 292-297. 1993.

[Lee97] M. T.-C. Lee. *"High-level test synthesis of digital VLSI circuits"*. Artech House, 1997.

[Man86] H. de Man, J. Rabaey, P. Six and L. Claesen. *"Cathedral II: A silicon compiler for digital signal processing"*. In IEEE Design and Test of Computers, pp. 13-25, Dec. 1986.

[Nie91] T. M. Niermann and J. Patel. "HITEC: A test generation package for sequential circuits". *Proc. European Design Automation Conference*, pp. 214-218,1991.

[Rab91] J. M. Rabaey, C. Chu, P. Hoang, and M. Potkonjak. *"Fast prototyping of datapath-intensive architectures"*. IEEE Design and Test of Computers, pp. 40-51, 1991.

[Sap84] S. Sapiro, R. Smith. *"Handbook of Design Automation"*. CAE System, Inc. 1984.

[Std90] 1149.1-1990 IEEE Standard Test Access Port and Boundary Scan Architecture.

10 ADVANCES IN ATPG BY EXPLOITING THE BEHAVIORAL VIEW

Walter Geisselhardt
Heinz-Dieter Huemmer

10.1 Introduction

In general three representations of a digital circuit or system can be distinguished,

- the structural at switch- or gate-level,

- the functional and

- the behavioral,

on base of which, in accordance with an appropriate fault model, test patterns are generated. **Hierarchical** representations encompass two or more of these levels.

Structural descriptions at gate level, assuming that the logic symbols for AND gates, OR gates and inverters are an accurate reflection of structure, as well as at switch level are in use. Switch level descriptions, are mostly used in conjunction with **hierarchical** test pattern generation.

Structural descriptions of circuits, however, are not always available and, even if available, control information is used additionally to reduce the expense of test pattern generation. Consequently, engineers are investigating ways to use functional and behavioral models in test and simulation. These models are described in HDLs, e.g. VERILOG, VHDL.

HDLs are called **non-procedural** languages, since they don't attach significance to the order of occurrence of statements. This is in contrast to **procedural** languages which impose an order on activities.

J.C. López et al. (eds.), Advanced Techniques for Embedded Systems Design and Test, 231-259.
© 1998 *Kluwer Academic Publishers.*

The flexibility of the languages permits to model a circuit at a level of detail ranging from structural to functional or behavioral. A design may even be expressed as a composite of elements, some of which are expressed in terms of structure and others expressed in terms of function and behavior. Difference between the disparate views are demonstrated on a 2- to -1multiplexer with control input C and signal inputs A and B in figure 10.1.

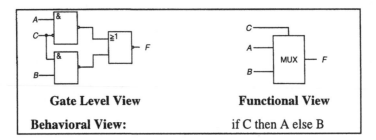

Gate Level View	Functional View
Behavioral View:	if C then A else B

Figure 10.1. General Views of a System

A complex circuit can be looked at as an interconnection of modules each performing a particular function. Modules consist of functional primitives such as ALUs, ADDERS, MULTIPLIERS, MUXs, COUNTERS, etc. which can be broken down into GATES and Transistors. This **hierarchical** view can be exploited in various ways for test pattern generation and for testing as it is for design.

10.2 Structural Test Generation, Definitions and Test Generation Approaches

10.2.1 Definitions

Structural Description

A structural description of a circuit or system gives a one-to-one correspondence between hardware components of the system and the descriptive elements. All operators represent hardware structures, the timing features of the system are established by clock signals or may be realized as event-driven.

Functional Description

The functional description of a circuit gives its input-output relation in terms of an expression out = f(in). Combinational circuits e.g. can be described by Boolean Functions.

Structural Test

A test is called structural if stimulus and response of the circuit is derived by its structural description in terms of a fault model assumption.

10.2.2 Basics of Structural Test Generation

Causes for circuits and systems to fail are defects of hardware mainly by improper manufacturing. Those defects become apparent during operation.

At system level defects manifest themselves by faulty behavior, as mal- or non-functioning, at hardware level they become visible as **faults** like

- signals stucking to 0/I,

- delayed switching or

- crosstalk.

So the idea of testing the hardware is to apply pre-calculated stimuli to the inputs and determine the output response of the circuit. If the circuit passes the test, it will (probably) operate correctly.

To derive the stimuli and to decide whether they are complete it has to be investigated, what hardware defects are likely and how they manifest themselves. This leads to the concept of **fault models** which represent the effect of defects by means of the changes they produce in the circuit signals.

Investigations revealed that most defects could be categorized as **Stuck-at Faults** but this is technology dependent. The stuck-at fault model assumes a signal line being permanently pulled to VCC (s-1 fault) or to GND (s-0 fault).

To make such faults apparent at the circuit's output the so called **path sensitization** is applied, i.e. to exercise any path between inputs and outputs sensitizing one at a time. If the output changes its value with the input, assume the hardware being involved is flawless.

The detection capability of this method is restricted to those faults not changing the structure, esp. shorts which, in fact, are not 'stuck-faults'. Hence, new fault models had to be created. The most popular are explained in short.

Stuck-open and **Stuck-on faults** are characteristic for CMOS-technology. If one of the complementary CMOS transistors in a gate by a defect permanently cuts off (s-open) or conducts (s-on) the result is a malfunction of the implemented gate. S-open faults even lead to undesired memory effects which only can be detected by sequences of stimuli.

Bridging faults manifest themselves as stuck-faults at a dominating signal value for bipolar technology. This doesn't hold for CMOS technology since there is no dominating signal value. Therefore, bridging faults have to be taken special care of, e.g. by inserting dummy gates on site of the short. IDDQ-testing seems to offer an easier way out.

Another undesirable effect of bridging faults in combinational circuits is a possible memory effect shown in figure 10.2. Like S-open faults in CMOS these effects only can be detected by sequences of stimuli.

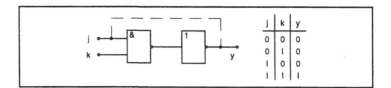

Figure 10.2. Bridging Fault Creating a Memory Effect

Delay faults might be caused by defects which don't manifest themselves as stuck-faults. This is especially the case in CMOS-circuits. To detect them it is necessary to propagate rising and falling edges along signal paths [Bra95].

Nevertheless, the prevailing fault model under which tests are developed is still the stuck-fault model. The faults covered by this model must observe the following restrictions:

- logic is altered,

- single faults,

- permanent,

- circuit's structure is unaltered.

Faults in **redundant parts** will remain undetected (figure 10.3).

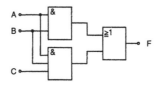

Figure 10.3. Circuit with Redundancy

With respect to **functional descriptions** there are two further methods to be mentioned here.

The traditional method of Boolean Differences is based on the Boolean function of a circuit for each output. It calculates the output dependency of every variable in the function and by this derives tests.

A modern method to express the Boolean function is by binary decision diagrams, BDDs. Akers [Ak78] proposed a test generation method based on BDDs.

The importance of BDDs should not be seen as a base for test generation itself but as a tool to support test generation and fault simulation. This will become clearer later in this chapter.

10.2.3 Algorithms for Combinational Circuits

The first true algorithm for ATPG on base of the circuit's structure and the stuck-fault model was the **D-Algorithm** [Roth66].

D stands for discrepancy and is a composite signal. It simultaneously denotes the signal value on the good machine (GM) and the faulty machine (FM) according to the table in figure 10.4.

GM	FM	
	0	I
0	0	D'
I	D	I

Figure 10.4. Table for Signal Values

Where the good and the faulty machine have the same value, the table will have a 0 or I. Where they have different values, there is a D in the table indicating the difference. D indicates a good value I and a faulty of 0 whereas D' indicates good 0 and faulty I.

The algorithm starts with a list of potential faults. Picking one of them the D-Algorithm tries to forward **propagate** the signal at the fault site to a primary output and backward **justify** the signals to the primary inputs. Propagation and justification are the two procedures which take place within the D-Algorithm.

The generated tests are devised for one fault at a time. Each test, however detects numerous additional faults. Which ones is determined by **fault simulation** with each newly derived test. Faults already covered are discarded from the fault list.

Since time complexity of fault simulation is approximately proportional to n^2, in contrast to test pattern generation which is proportional to 2^n, fault simulation became an important pillar of ATPG and an object of research itself, e.g. [GM93, Mo94]. Several fault simulators have been developed, e.g. PROOFS [CP90] or PARIS [GK91b], and implemented in the test generation procedures. Figure 10.5 shows the flow of the simulation based test generation.

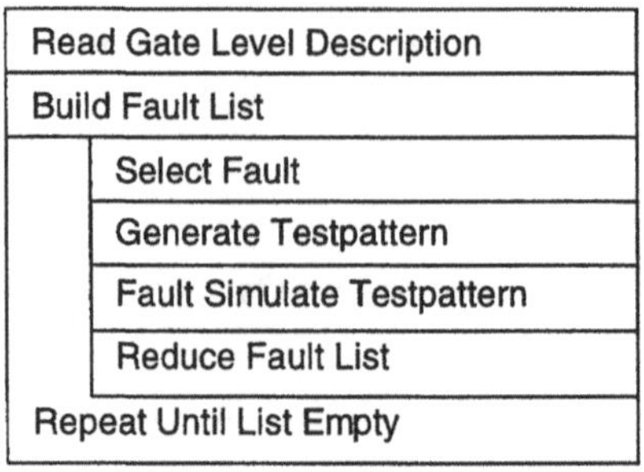

Read Gate Level Description
Build Fault List
Select Fault
Generate Testpattern
Fault Simulate Testpattern
Reduce Fault List
Repeat Until List Empty

Figure 10.5. Flow of Simulation Based Test Generation

The number of stuck-faults to be considered can be greatly reduced by virtue of fault **equivalence** and fault **dominance**.

Faults are called equivalent, if there is no logic test that can distinguish between them. Fault a is said to dominate fault b, if any test which detects fault b will detect fault a. A test pattern which detects one member of a group of equivalent faults automatically detects the entire group. The same holds for dominated faults. The procedure of building groups of faults, **fault classes**, which are detected by the same pattern is called **fault collapsing**.

Another important fact to realize is the following. When assignments are made to individual gates, these assignments often possess **implications** beyond the gate being processed. Depending on a circuit's structure several of the succeeding signal lines might be forced to certain values as consequence of an assignment. The same holds for preceding signal lines. Assignment of a 0 to the output of a NAND gate e.g. implies that its inputs are 1. These implications can be determined, when an assignment is made to save computation time.

Powerful ATPG systems provide procedures to **learn** implications from the circuit's structure and apply the knowledge to further steps.

In circuits that rely heavily on reconvergent fan-out, as is mostly the case, the D-Algorithm can become mired in large numbers of conflicts. They result from the justification process which, proceeding back towards primary inputs along two or more paths, frequently arrives at a point where signals converge with conflicting requirements.

As a result, the D-Algorithm must find a node where an arbitrary decision was made and choose an alternative assignment. The amount of time for that can become prohibitively large.

The well known **PODEM** test generation algorithm [Goe81] addresses the problem of large numbers of remade decisions by selecting a fault and then working directly at the inputs to create a test. PODEM begins by assigning Xs (unknown) to all inputs. It then starts assigning arbitrary values to primary inputs. Implications of the assignments are propagated forward. There are two propositions which control the procedure:

1. The signal for the stuck-fault being tested has the same logic level as the stuck-fault.

2. There is no signal path from an internal signal net to a primary output such that the net has value D or D' and all other nets on the signal path are at X.

If either of them is true, the assignment is rejected. This employs a so called **branch-and-bound** method and introduces the idea of test pattern generation as a **search process** (see figure 10.6).

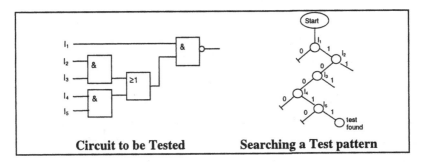

| **Circuit to be Tested** | **Searching a Test pattern** |

Figure 10.6. Test Generation as a Search Process

The ease or, better, difficulty to find a test for a particular fault is expressed by its **testability**. Testability is a composite of **controllability** and **observability**. Controllability measures the ease to assign a 0 or I to the site of the stuck-fault, observability values the ease to propagate the resulting signal to an observable output.

Testability measures are used to decide

*   which fault to select next for test pattern generation, hoping that pattern for easy to test faults will also detect the difficult ones,

*   which input or path to choose in order to control or propagate a signal.

They can speed-up test pattern generation considerably. Numerous testability measures were proposed and are in use with the various ATPG systems, e.g. SCOAP, CAMELOT, COP, LEVEL.

10.2.4 Algorithms for Sequential Circuits

Test pattern generation for combinational circuits is known to be NP-complete. Since the problem is even more complicated with sequential circuits, the only way to cope with circuits of significant size is to heavily rely on heuristics.

Sequential circuits here are restricted to **synchronous circuits** according to the Huffman model. In sequential circuits a test for a particular stuck-fault consists of a sequence of patterns and clock pulses applied to the primary inputs to stimulate the fault and propagate it to an output where it can be observed. The sequence of states the circuit is stepped through can be represented by an equal number of consecutive copies of the circuit. Strictly speaking, one copy for every clock cycle, so called time frames, figure 10.7.

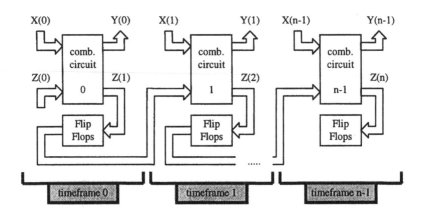

Figure 10.7. Iteration of Combinational Part in Sequential Circuits

Figure 10.7 illustrates the inputs $X(i)$, the outputs $Y(i)$ and the state vector $Z(i)$ for the specific time frame i. If the flip-flops in such iterations are viewed as interfaces the whole structure can be handled as an overall combinational circuit. The signal vectors $Z(i)$ feeding the iteration i are called secondary inputs in contrast to the primary inputs $X(i)$. It is obvious that their settings depend on the previous iterations, hence, one of the most important problems is to find out initial states $Z(0)$ and ways to drive the circuit into those states.

Standard procedure for sequential test pattern generation follows the **reverse time processing (RTP)** approach. It begins with the 'last' time frame, the final step of the fault propagation procedure. If the primary and secondary inputs of this time frame are assigned and justified, the algorithm proceeds to the previous time frame and so forth.

Iteratively working backward in time, one strives to attribute secondary inputs to primary inputs step by step. A number of algorithms have been proposed and implemented based on the RTP approach. They differ in how the propagation path is determined.

One of the most successful algorithms is the BACK-Algorithm [Che88] based on which a number of further improved algorithms were developed and implemented, e.g. DUST [Go91, GK91a] and FOGBUSTER [Gl94].

They all include improved heuristics to reduce the search space and the ability to **learn** unreachable states of the circuit according to values the secondary inputs (flip-flops) cannot take on.

10.2.5 Complexity Limits

As already mentioned the test generation problem for combinational circuits is NP-complete, i.e. in worst case the generation of tests for a circuit with n components consumes time that grows by the power of n. This can be reduced, as shown before, by implementation of some intelligent methods like fault simulation and learning to reduce the number of searches.

Complexity of test generation even increases more drastically with sequential circuits. Figure 10.7 illustrates that the combinational part of those circuits is iterated in time frames. Hence, the test generation complexity of the combinational part is at least multiplied by the number of iterations.

As shown by some examples pure structure oriented test generation schemes although highly sophisticated cannot keep pace with the increasing complexity of circuits and systems. Consequently, new methods have been invented which are not completely dependent on the structure but on the functional or behavioral description of a circuit.

10.3 Behavioral Test Generation, Definitions and Test Generation Approaches

10.3.1 Definitions

Behavioral description

A behavioral description of a circuit expresses how the outputs react on input stimuli by means of a hardware description language (HDL) independently of the internal structure.

Behavioral test

A test is called behavioral if stimulus and response of a circuit is derived from its behavioral description in terms of a fault model or some other completeness criteria.

10.3.2 Basics of Behavioral Test Generation

Although there are some approaches to behavioral test generation in the 1980th e.g. [Jo79, BaAr86, GMo89] the really applicable schemes came up in the beginning of the 1990th e.g. [BM89, NJA90, GSC91, HVT91].

Lack of the first approaches in behavioral test generation was a generally accepted HDL. To generate tests one had to produce a behavioral description of a circuit following its design. This was not acceptable.

After the HDL Verilog had been extended to behavior descriptions and VHDL (very high speed integrated circuit hardware description language) had been standardized by the IEEE [IEEE92] both became integrated into the design process. Nowadays, all circuits and systems are designed starting with a behavioral model mainly in Verilog HDL or VHDL which can be simulated. Further design steps down to silicon are derived from these descriptions. In the beginning these derivations have been done mostly by hand but there is ongoing work to automate this process. Nevertheless the fiction of a fully automatic silicon compiler fades at the moment. Taking into account a top-down design scheme like this it makes much more sense to derive tests from behavioral HDL descriptions.

The description level closest to the structure is the register transfer level (RTL). Most of the behavioral test generation approaches are based on this level.

Again, there are fault models with fault lists which the tests have to cover. According to structural test methods the faults assumed are stuck-at data or wrong assignments for the data flow or stuck-THEN / stuck-ELSE for the control flow.

Another method stems from ideas of structural software test, i.e. to cover all control paths, all branches or all assignments of the description by a test pattern set.

Although these tests, applied at speed, are apt to detect any fault affecting the circuit's function including timing faults, they are not widely accepted, since there is no proof of what defects they actually cover. Recent investigations, however, show a stupendous reproduction of gate-level faults by a special RTL model [YoSi96].

After all, there are no commercial behavioral test generation tools available, up to now.

10.3.3 Algorithms for Behavioral Test Generation

To represent the wide area in this field some characteristic approaches are explained in short.

One of the oldest approaches for RTL descriptions is the **S-Algorithm** [Li84]. This algorithm assumes 9 fault types that result from structural fault assumptions with respect to their effect on the RTL description. By using symbolic simulation (S-Algorithm) of the RTL description tests are derived to cover the fault list assumed. Some derivatives of this approach have been published later on.

If going up to higher levels of description the relation of structural faults to their effect on the language constructs becomes blurred. Hence, the fault assumptions have to come from the language constructs themselves. In these approaches **the language constructs are perturbed** by stuck-at faults at inputs, outputs or state variables, by control faults and/or by general function faults.

The first proposal concerning perturbation was done by Levendel and Mennon [LM82], but at that time no HDL has been generally accepted. The authors apply a poor HDL for describing the circuit's behavior. For test generation the basics of the D-Algorithm are applied for propagation and justification of D values through the HDL flow graph.

James R. Armstrong and his group were one of the first applying **perturbation schemes to VHDL** descriptions to develop the behavioral test generator BTG [BaAr86, NJA89] and [ChAr94]. Again language constructs are perturbed and propagation and justification are calculated for tests.

Various derivatives of this scheme can be found, too, which are not mentioned in detail here.

The third and highly actual test generation scheme based on behavioral descriptions makes use of structural software test techniques [My79]. Two groups and their tools are to be mentioned, Giambiasi and Santucci [GSC91] on one hand and Geisselhardt [GMo89], [HVT91] and [Ve92] on the other. Both schemes make use of the VHDL behavioral description's control and data flow.

Giambiasi and Santucci derive **two internal models** from the VHDL description. The control model of the inherent **control in terms of Petri nets** and a data model with representation of **data and linked operations**. A behavioral fault model is defined with respect to the internal models including faulty operand or operation, wrong control path and skipped operations. Similar to the above mentioned perturbations faults are injected into the models (manifestation). The injected faults are sensitized, propagated and justified by exploiting the models. The scheme also makes use of a behavioral fault simulator to reduce the fault list.

The FUNTEST approach [GMo89] was based on SMILE/FBDL descriptions but had a similar approach in covering a behavioral fault list.

FunTestIC, finally, [Ve92] (**Fun**ctional **Test generator** for ICs) approaches the behavioral test generation completely different. It doesn't aim at covering a given fault set but **explores the data flow**. The idea is that only those patterns are tests which drive certain data from inputs to the outputs sometimes taking detours over memorizing elements (variables or registers). The data paths and the associated operations are governed by the control flow. Hence, FunTestIC derives an **internal model** from the VHDL description comprising the control and related data flow in terms of PROLOG facts. A special graph search algorithm extracts all input-output paths, so called **test cases**, from the data flow. The core algorithm of FunTestIC then solves the problem of driving data from certain inputs to certain outputs, i.e. **calculate symbolically** the stimuli and the response. The stimuli are expressed in symbols, i.e. there is a grade of freedom for choosing definite values. The response is given in functional representation and has to be calculated for given stimuli. Hence, a so called test frame compresses the information to a set of data being driven by identical control signals from input to output. The data itself can be chosen randomly. The completeness criteria for FunTestIC to generate test frames is

- to cover all data paths of the model at least once and keep track of associated control path coverage
- if a control path is not covered find a covering test frame.

The general operation mode of the FunTestIC system is shown in figure 10.8.

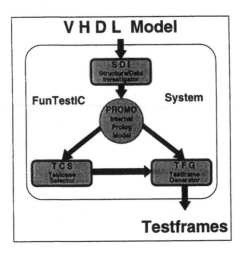

Figure 10.8. The FunTestIC System

Although the test sets, like those of all other behavioral test generation approaches, are not generally accepted, FunTestIC has other valuable features, e.g. it can generate frames to reach predefined nodes or node sequences in the control flow. This feature will be very helpful as shown further.

10.3.4 The Gap between Behavior and Structure

It has already been mentioned before that the design process is going to be automated i.e. the behavioral description is finally converted (compiled) to physical structures. If there is a fully automatic conversion process with definite relations it's no problem to link the effects of physical failures to the high level description constructs. In practice, however,

- a lot of conversion steps are done by hand in most cases and
- after conversion to structure there are a lot of optimization processes blurring the clear relation.

On the other hand, for test generation, particularly for approaches dealing with behavioral information of the circuit, it is of great importance to have knowledge about the behavior to structure relations. The rapidly growing complexity of circuits enforces new test generation approaches to master the complexity of structural test generation. This task brings together behavioral and structural descriptions for test generation.

10.4 Mastering the Complexity of Test Generation

10.4.1 Hierarchical Blindness Problem

It has been shown that behavioral test generation works without knowledge about the details of the circuit's structure. Structural test generation on the other hand operates unaware of the circuit's function / behavior. This phenomenon is termed **hierarchy blindness problem** by Lee and Patel [LiPa91].

The behavioral level is blind because it cannot observe structural details. An adder e.g. realizing the assignment C := A + B in a behavioral description can lead to different structures:

- It can be a counter loaded by the value of A counted upwards value of B times,

- it can be a serial adder, adding digit by digit in timesteps or

- it can be a well known parallel adder with or without a ripple carry.

```
M := A;
for i in 1 to B loop
  M := M +1;
end loop;
C <= M;
```

Algorithm a

A	- variable
A(i)	- digit of variable
base	- base of number- system

```
carry := 0;
for i in 0 to (n-1) loop
  M(i) := A(i) + B(i) + carry;
  if M(i) >= base then
    M(i) := M(i) - base;
    carry := 1;
  else carry := 0;
  end if;
end loop;
M(n) := carry;
C <= M;
```

Algorithm b

Figure 10.9. Different Approaches for Assignment C := A + B

Each of these realizations is performing the addition but behaves differently and, hence, will be tested in a different way. Only when the behavioral description gets precise enough (see figure 10.9) the blindness is surmounted.

On the other hand structural descriptions of a circuit are blind towards it's function rsp. behavior. If a structural block is embedded and its surrounding constrains the inputs of this block, this will never be recognized by e.g. a structural test generator. Lee and Patel term this **global functional constraints** [LiPa92]. Figure 10.10 gives their example of an embedded macro to distinguish control from bus constraints. A control constraint is meant by selecting only 4 of 64 functions of the ALU. A so called bus constraint is given by limiting one adder input to the value 15.

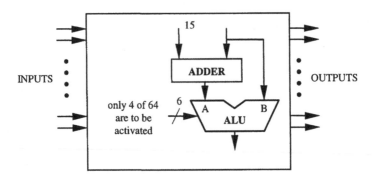

Figure 10.10. Example of Global Functional Constraints

A structural test generation tool deriving tests for this macro will never get this information on global constraints.

10.4.2 Exploiting Behavioral Information to Guide Structural Test Generation

The blindness problem for structural test generation tools arises mainly from control lines. This holds for functional selections not globally covered, like shown in figure 10.10, or even more worse for state sequences with perhaps unreachable states in sequential logic. Structural test generators spend a lot of effort until at the end the justification process fails in those cases.

Some papers deal with overcoming this problem from the structural point of view. They all make use of control and data flow information how to reach a particular state of a circuit, which is available from its higher level descriptions, to speed up the structural test generation process e.g. [LiPa92, Sa89] and [ViAb92]. The way they explore the behavioral information differs.

The FunTestIC approach extracts the relevant information from VHDL and calculates test frames as constraints for the structural test generator. Preliminary to this procedure is a description with two special properties:

- The schedule of the HDL model and the circuit to be tested are the same.

- The allocation of operations in the HDL model to the functional macros in the circuit has to be known.

To clear these properties a short look is taken on the principles of high level design using an example from [ArGr93].

The VHDL code to derive the outputs X and Y from inputs A, B, C and D may be by two concurrent signal assignments. Figure 10.11 also gives the corresponding data flow graph DFG.

$$X <= E * (A + B + C)$$
$$Y <= (A + C) * (C + D)$$

VHDL Code

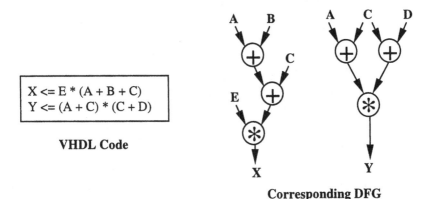

Corresponding DFG

Figure 10.11. Example of Circuit Design

Among the various possibilities to construct a circuit performing these operations there are two extreme solutions:

1. In space domain, i.e. the circuit consists of 4 adders and 2 multipliers. As the two concurrent signal assignments in VHDL characterize, this is done with the delay time of the longest path which can be seen as one minimal timestep.

2. In time domain, i.e. the circuit consists of one functional unit which performs the add and multiply operation or two different units adder A1 and multiplier M1. Hence, registers have to be included to store intermediate results. Now, in time domain with only A1 and M1, the calculation will need 5 timesteps (control steps) but smaller delay for each timestep.

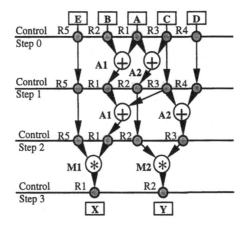

Figure 10.12. Scheduled DFG

The high level design task is to find an optimal distribution of functional units to operations (allocation) combined with an optimal / minimal number of timesteps (scheduling).

One possible result of scheduling and allocation for the function above with respect to some optimization criteria is drawn from [ArGr93] and given in the DFG of figure 10.12. At each control step the registers to store the intermediate results are drawn.

This scheduled DFG can be transformed to a behavioral VHDL description. Figure 10.13 shows one possible VHDL single-process model. Note that the assignments are numbered at the end of each line for reference.

```
entity EXAMPLE is
  port (A, B, C, D, E: in INTEGER;
        CLK, RST: in BIT;
        X, Y: out INTEGER);
  end EXAMPLE;
architecture MRTL of EXAMPLE is
  variable STEP, R1, R2, R3, R4, R5: INTEGER;
begin
  XMP: process (CLK)  begin
  if RST = '1'  then                              -- 0
    STEP := 0;                                    -- 1
  elsif CLK'EVENT and CLK = '1' then              -- 2
    case STEP is
      when 0 => R1:=A; R2:=C; R3:=D, R4:=B; R5:=E;   -- 3, 4 to 8
                X <= R1; Y<= R2; STEP := STEP+1;      -- 9, 10, 11
      when 1 => R1 := R1 + R2; R2 := R1 + R3;         -- 12, 13, 14
                X <= R1; Y <= R2; STEP := STEP+1;      -- 15 to 17
      when 2 => R1 := R1 + R3; R3 := R3 + R4;         -- 18, 19, 20
                X <= R1; STEP := STEP+1;               -- 21, 22
      when 3 => R1 := R1 * R5; R2 := R2 * R3;         -- 23, 24, 25
                X <= R1; Y <= R2; STEP := 0;           -- 26, 27, 28
    end case;
    end if;
  end process XMP;
```

Figure 10.13. VHDL Model for the Scheduled DFG

Adder A1 maps to statements 13 and 19, A2 to 14 and 20, multiplier M1 to 24 and M2 to 25. These relations often are termed **synthesis links**.

VHDL descriptions of this style can be synthesized by e.g. the Synopsys Design Compiler. The circuit derived will consist of 5 registers for data storage, 4 functional units A1, A2 for addition and M1, M2 for multiplication and a finite state machine performing the 4 control steps. Figure 10.14 shows the derived data path.

Let's assume, a structural test generator should generate tests for this circuit and, following the single fault assumption, it will suppose a stuck-fault e.g. in unit M1. First, the fault propagation task will forward propagate the fault. This will be done through the MUX at input A and R1 to output X and, hence, needs one timestep. To control this fault a backward search has to start which will make no problems via R5 and input E but will need a lot of effort because the data loops M1→R1→MUX→M1 or M1→R1→MUX→A1→R1 etc. may force numerous backtracks.

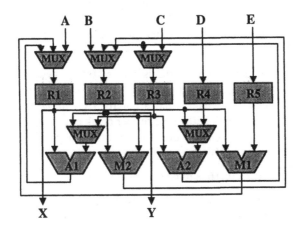

Figure 10.14. Data Path to Synthesize

At this point the behavioral information and the principle of FunTestIC can support the process and, consequently, omit the backtracks. From the allocation it is known that the unit M1 maps to assignment number 24 (see figure 10.13). FunTestIC's internal model PROMO (figure 10.8) reflects the control flow on one hand and the so called data dependency graph DDG on the other. Figure 10.15 shows the DDG of the example. Note that each directed arch reflecting a dependence is labeled by the number of statement(s) corresponding to it.

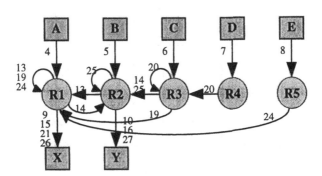

Figure 10.15. Data Dependency Graph

The task of the **test case selection** part (TCS) of FunTestIC (figure 10.8) is to derive a control node (statement-) sequence to access node 24. Archs labled 24 in the DDG are those from R5 to R1 and R1 to R1, thus, reflecting the assignment R1

:= R1*R5. Further, paths from the inputs have to be found. Statements 4 and 8 are those corresponding to archs A to R1 and E to R5. The statement numbers are sorted in increasing order giving [4-8-24] as a **goal to control** statement 24. From the DDG it can also be seen that observation can be done via output X. One of the possible statements 9, 15, 21 or 26 can be a candidate. The increasing order in the statement numbering scheme tells that the number has to be greater than 24, i.e. here number 26. Hence, the **goal to control and observe** 24 ist [4-8-24-26], a complete **test case**.

The most complex part of FunTestIC is the **testframe generator** (TFG). The TFG gets the goal for frame generation as a list of control nodes like shown above. By control flow analysis, i.e. try to find a control path through the flow and, if necessary, step backwards in time (assume further clock edges) it reaches the goal to find frames to control node 24 and to observe the result at the outputs. The testframe derived is shown in figure 10.16.

The testframe in figure 10.16 is to be read as follows:

- Steps are pattern sequences in which inputs change their value. In case of clocks being marked with a **P** the step includes a clock pulse.

- Fields marked with **x** mean that a value is undefined and not of relevance for the test goal.

- Values in the first position marked with **$** are symbolic data, i.e. they can be determined within their given limits, e.g. 8-bit values ranging from 0 to 255.

| Step | INPUTS | | | | | | INTERNAL ELEM. / OUTPUTS | | | | |
no	RS	CK	E	B	A	C	D	R5	R2 / Y	R1 / X	R3	R4
1	1	x	x	x	x	x	x	x	x	x	x	x
2	0	P	$D5	$D2	$D1	$D3	x	$D5	$D2	$D1	$D3	x
3	0	P	x	x	x	x	x	$D5	$D1 +$D3	$D1 +$D2	$D3	x
4	0	P	x	x	x	x	x	$D5	$D1 +$D3	$D1 +$D2 +$D3	x	x
5	0	P	x	x	x	x	x	$D5	x	($D1 +$D2 +$D3) *$D5	x	x

Figure 10.16. Testframe for M1

The testframe shows symbolically which data is relevant at which timestep. To test the unit M1, the symbolic values $D1, $D2 and $D3 have to be added as one input of this unit and $D5 as the other. All values have to be set to the inputs for the first clock cycle after reset and 2 cycles have to follow. In this case all 4 values can be chosen without constraints. The 3rd clock cycle makes the output of unit M1 observable at the primary output X through R1. Within this given frame a structural test generator can generate tests for the functional unit M1.

It should be mentioned that FunTestIC is not limited to integer calculations. There are powerful algorithms implemented as well based on binary decision diagrams to solve arithmetic calculations - all IEEE 1076 standard functions are implemented - and to solve path predicates for control flow paths by so called high level BDDs [EK95].

10.4.3 The Hierarchical Approach

The design example above showed how test generation can be supported if the structural modules link to statements embedded in a behavioral description. A somewhat different approach is for circuits with structural hierarchy in which the modules are described in synthesizable VHDL code. As an example [KE96], figure 10.17 depicts an industrial telecommunication circuit composed of several modules of which the ENCODER is assumed to be isolated and tested without any constraints.

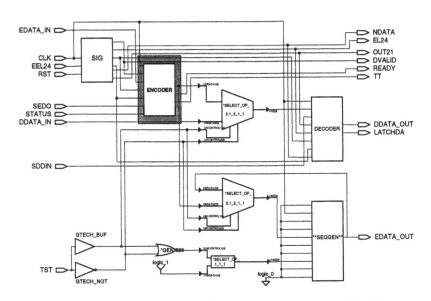

Figure 10.17. Example of a typical Telecommunication Circuit

The ENCODER module itself is made up of two submodules, CONTROL and ENC, like shown in figure 10.18.

The CONTROL submodule is described in VHDL by two cooperating processes, SYNCH and COMBIN. Interaction of the two processes is shown in figure 10.19.

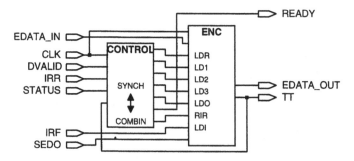

Figure 10.18. Decomposition of the ENCODER module

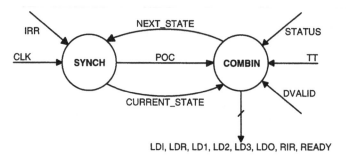

Figure 10.19. Communication of Processes SYNCH and COMBIN

```
SYNCH: process
 begin
 wait on CLK;
 if IRR = '1' then
   CURRENT_STATE <= NEXT_STATE;
   case NEXT_STATE is
     when "000" => POC <= "00";
 ...
 ...
     when others => NULL;
   end case;
 else
   CURRENT_STATE <= "000";
   POC <= "00";
 end if;
end process SYNCH;
```

```
COMBIN: process
 begin
 wait on CURRENT_STATE,
         DVALID,STATUS, TT;
 case CURRENT_STATE is
   when "000" =>
     LDI <= '0';
     LDR <= '0';
     LD1 <= '0';
     LD2 <= '0';
     LD3 <= '0';
     LDO <= '0';
     READY <= '0';
     RIR <= '0';
     if (STATUS = '0' or DVALID = '0') then
       NEXT_STATE <= "000";
     else
       NEXT_STATE <= "001";
     end if;
   when "001" =>
 ...
 ...
 ...
end process COMBIN;
```

Figure 10.20. Extractions of the VHDL Code for Submodule CONTROL

The VHDL models in this case follow some design rules because they are tailored to be compiled by the Synopsys Design Compiler. Like the names reveal, the CONTROL module is a finite state machine which will be built by memory elements, described in submodule SYNCH, and by combinational functions for the next state and the output, described in submodule COMBIN. Figure 10.20 gives extractions of the VHDL code to get an impression of the description style.

In figure 10.20 at the left, process SYNCH waits on a clock event to store the current state from NEXT_STATE being input from process COMBIN (figure 10.20 at the right). COMBIN is triggered by signal CURRENT_STATE and some other signals and determines the next state and the output.

This two process VHDL model is given to FunTestIC to generate testframes for all input-output paths. The testframes derived for the process CONTROL are a set of pattern sequences which can always be reproduced at the module's outputs. Figure 10.21 gives one oft the testframes derived by FunTestIC. Note that each pattern has two steps because of the two processes corresponding with each other.

The first line in figure 10.21 gives the type of the signals and line 2 the names. The lines 3 to 13 depict the signal values to be input, the output signals produced and the states reached. An **X** denotes undefined values which become defined later and a **D** denotes input values of no further relevance for this frame.

sig. type	In	in	In	in	in	out	out	out	out	out	out	out	out	state	state	state
Pat.(-2 , [	Clk,	irr,	Dva lid,	sta tus,	tt,	ldi,	ldr,	ld1,	ld2,	ld3,	ldo,	rir,	read y	cur_ state	next_ state	poc]).
Pat.(0.0 ,[	"1",	"0",	"0",	"0",	"0",	"x",	"x",	"x",	"x",	"x",	"x",	"x",	"x",	"xxx",	"xxx",	"xx"]).
Pat.(1.0 ,[	"0",	"0",	"0",	"0",	"0",	"x",	"x",	"x",	"x",	"x",	"x",	"x",	"x",	"xxx",	"xxx",	"xx"]).
Pat.(1.1 ,[	"0",	"0",	"0",	"0",	"0",	"x",	"x",	"x",	"x",	"x",	"x",	"x",	"x",	"000",	"xxx",	"00"]).
Pat.(2.0 ,[	"1",	"0",	"1",	"0",	"0",	"x",	"x",	"x",	"x",	"x",	"x",	"x",	"x",	"000",	"xxx",	"00"]).
Pat.(2.1 ,[	"1",	"0",	"1",	"0",	"0",	"0",	"0",	"0",	"0",	"0",	"0",	"0",	"0",	"000",	"000",	"00"]).
Pat.(3.0 ,[	"0",	"1",	"1",	"1",	"0",	"0",	"0",	"0",	"0",	"0",	"0",	"0",	"0",	"000",	"000",	"00"]).
Pat.(3.1 ,[	"0",	"1",	"1",	"1",	"0",	"0",	"0",	"0",	"0",	"0",	"0",	"0",	"0",	"000",	"001",	"00"]).
Pat.(4.0 ,[	"1",	"1",	"0",	"D",	"0",	"0",	"0",	"0",	"0",	"0",	"0",	"0",	"0",	"000",	"001",	"00"]).
Pat.(4.1 ,[	"1",	"1",	"0",	"D",	"0",	"0",	"0",	"0",	"0",	"0",	"0",	"0",	"0",	"001",	"000",	"00"]).
Pat.(5.0 ,[	"D"	"D"	"1",	"D",	"0",	"0",	"0",	"0",	"0",	"0",	"0",	"0",	"0",	"001",	"000",	"00"]).
Pat.(5.1 ,[	"D"	"D"	"1",	"D",	"0",	"1",	"0",	"0",	"0",	"0",	"0",	"0",	"0",	"001",	"001",	"00"]).

Figure 10.21. Test frame derived from Process CONTROL

The structural sequential test pattern generator DUST [GK91a] is able to generate patterns under input constraints. Frames like shown in figure 10.21 can serve as input constraints for following modules. In the example DUST starts structural test pattern generation for the module ENC from its gate level description. In the frame of the patterns given from CONTROL, DUST computes the input data for the signals EDATA_IN, IRF and SEDO to reach its goal to detect a stuck-fault assumed.

A short look to some results of this work. The ENCODER module is proved to be extremely untestable. A commercial tool wound up with a mere 4.1 % stuck-fault coverage for the non-scan version of the module itself. With DUST, fault coverage went up to 30.33%, the longest pattern sequence comprising 85 patterns.

Putting 42 of the total of 105 flip-flops into a scan-path, DUST reached a fault coverage of 98.8% (at 99.39 % efficiency) with 11.029 test pattern, the number of patterns in a sequence being 150 on average. The longest sequence comprised 359 patterns.

Now, considering the ENCODER as two modules and the test generation procedure like shown above.

For the ENC module itself, provided with a partial scan including 42 of its total of 100 flip-flops, DUST reached 99.81% fault coverage with test sequences of up to 61 patterns. Provided that CONTROL is able to deliver pattern sequences of that length, the same fault coverage would be reached if ENC and CONTROL were put together; if not, numerous faults within ENC would be reported untestable.

This is exactly what happens when CONTROL and ENC are put together as with the real design: Searching CONTROL' s VHDL description, FUNTESTIC found testframes of up to 13 patterns length. This means, CONTROL cannot provide sequences longer than 13 patterns. Under these constraints, DUST was able to produce test patterns of 81.13% fault coverage only, this, however, with remarkable 99.84 % efficiency.

Another example of a successful link between structural and behavioral test approaches is hierarchical test of embedded processors by self-test programs pursued in the STAR-DUST project [BiKa96]. The STAR-DUST approach is based on RESTART (retargetable compilation of self-test programs using constraint logic programming) a tool able to generate full self-test programs [BM95]. RESTART is the very first tool which has incorporated ideas employed in retargetable compilers (see [MG95] for survey). The other constituent is DUST [GK91a], the above mentioned constrainable structural test pattern generator.

RESTART automatically generates comprehensive processor test programs based on realistic fault models. This requires the availability of both RT-level and gate-level models for the same processor. These models are used by the tools implementing the essential tasks of the current approach:

1. synthesis of gate-level descriptions for each of the RT-level components,

2. test pattern generation for each of the RT-level components from their gate-level descriptions by the DUST system,

3. generation of machine instructions implementing justification and response propagation for each of the test patterns by the RESTART compiler,

4. fault simulation computing the coverage achieved by running self-test programs.

The procedure starts with a RT-level description of the processor under test. Input formats currently available for this purpose include VHDL or MIMOLA [BBH94]. From this specification an equivalent gate level description for each of the RT-level components is obtained by the SYNOPSYS logic synthesis tools.

Next, RT-level components are selected in a heuristic order and **for each** component the following steps are executed:

1. DUST **generates** structural **test pattern sets** for the component selected.

Constraints for input, output and state are taken into account. A general fault list is covered by simulation of all newly generated tests for a component. The generated test patterns are translated into Test Code Language (TCL) as input to the RESTART compiler.

2. The TCL-coded test patterns are **implemented into self test programs** by RESTART. The self test programs are to control faults in a component through primary inputs and to make them observable. A fault is observable by a jump of the microprogram control to an error label.

The following requirements must be met by processors to be tested with code generated by RESTART:

- The processor must be programmable.

- The processor to be tested must be able to perform a compare operation and a conditional jump.

- The program counter must be observable.

- A single instruction cycle is required.

3. The self test programs are used as initial stimuli for **fault simulation at the gate-level**. The test coverage is derived to reduce the global fault list.

There are two fault simulators available. FAUST (**Fau**lt **S**imulation **T**ool), a very efficient single pattern, single fault propagation simulator and PARIS (**Pa**rallel **I**terative **S**imulator), a parallel pattern single fault propagation simulator [GM93, GK91b]. The simulators assume microoperations in the microprogram memory as fault free.

RESTART, DUST and **FAUST** are the essential constituents of this test generation process, giving this approach its name STAR-DUST.

This approach has been demonstrated for two processors. First a small processor named SIMPLECPU shown in figure 10.22.

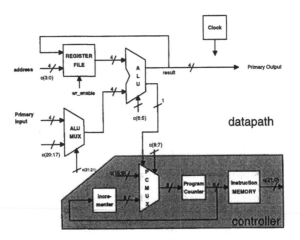

Figure 10.22. SIMPLECPU

The RT-level description of SIMPLECPU has been synthesized. The resulting circuit consists of 467 gates and 72 D-flip-flops.

First, structural test patterns for the register file were derived by DUST, taking into account some functional constraints. 193 test patterns for the register file have been generated. Fault simulation of the processor with these tests additionally achieved a fault coverage of 100% for the multiplexer ALUMUX and 49.08% for the ALU. To detect the remaining faults of ALU and ALUMUX, DUST, in the next iteration cycle, generated 17 test patterns. The resulting self test code comprises 51 instructions, i.e., three instructions to apply one ALU test pattern and to check the test response.

The CONTROLLER test is mainly done implicitly by the self-test program so far. Adding six jump instructions (conditional and unconditional) to check the remaining operations results in a fault coverage of 98.36% for the CONTROLLER.

The entire self-test program for SIMPLECPU comprises 250 microinstructions with a stuck-fault coverage (including data path and controller) of 99.63%.

Figure 10.23 illustrates the results in detail.

component	gates	DFFs	faults	det. faults	un tstb	fnct. unt.	abor ted	fault co- verage %	effic. %
register file	316	64	2,256	2,252	0	4	0	99.82	100
ALUMUX	13	0	76	76	0	0	0	100.00	100
ALU	70	0	436	436	0	0	0	100.00	100
controller	68	8	488	480	6	2	0	98.36	100
SIMPLECPU	467	72	3,256	3,244	6	6	0	99.63	100

Figure 10.23. SIMPLECPU Results

As a study for a more complex processor within the Belsign-Group STAR-DUST was applied to a SPARC processor [Sprc92] with a typical RISC architecture. The instruction set of this RISC processor is reduced to 32-bit words. The INTEGER UNIT of the SPARC fulfills the above mentioned preliminary of single-cycle instructions solely. Figure 10.24 shows a function diagram of the INTEGER UNIT.

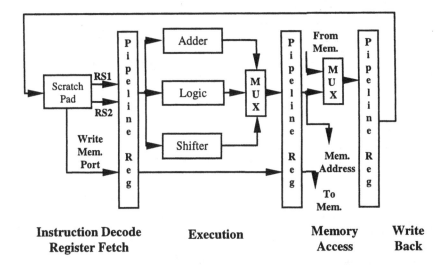

Figure 10.24. Integer Unit of the SPARC Processor

From the test generation point of view the pipeline architecture is of great interest. It can easily be seen from figure 10.24 that the pipeline registers hinder the test access to the module under test. Another important fact is known as functional hazard of pipelines. If a pipeline stage performs an operation that depends on a result just being evaluated by another stage, the operation has to be delayed. Hence, in those situations the self-test program of RESTART had to be rearranged a little bit by hand.

The synthesis of the INTEGER UNIT from RT-level comprises 5,239 gates with 31,596 stuck-faults which can be collapsed to 13,878 fault classes.

As shown above the test generation procedure is an iteration by choosing a new component as generation candidate and trying to cover the complete fault list by the test patterns just generated. The table in figure 10.25 shows how the test generation process developed.

The results show that the self-test program approach linking behavioral and structural test pattern generation can achieve high fault coverage for the data path as well as for the controllers.

In general, the hierarchical test approach divides the complex test problem for the entire processor in several less complex subproblems. Therefore, tests can be generated within remarkable shorter times.

component	# test patterns	# accum. testprogr. instr.	overall fault coverage %
adder 1. step	132	805	52.22
adder 2. step	7	853	52.23
shifter	45	1,117	70.96
logic compare	46	1,393	78.00
scratchpad 1. step	267	2,009	86.81
scratchpad 2. step	166	2,202	96.48

Figure 10.25. Test Generation Evolution for INTEGER UNIT

10.5 Concluding Remarks

As a short summary and for a close look to trends with respect to the sections before, some considerations.

It has been stated that structural test generation is at least NP-complete and with reality of designs impractical if there is no additional support to reduce the calculation effort. On the other hand no general and serious attack has taken place against the stuck-fault model, besides some exceptions for CMOS-logic. Test coverage nearly always is given with respect to stuck-faults.

Nearly all IC manufacturers make use of structural tests and reduce the test generation costs by implementing scan paths [WiPa82], i.e. chaining all flip-flops, hence, cutting or reducing the sequentiality of a circuit. This principle in other approaches is combined with built-in self test features by LFSRs (linear feedback shift registers) or MISRs (multiple input signature registers) for on-chip pseudo random test generation and test response compression [Cha77, HLM83]. The serious disadvantages of those approaches are extremely high test time, not at speed and not at function test. Generally speaking, the complexity of sequential test generation here is mastered by governing the inherent sequentiality with a tricky mechanism in using the state register for test access. The complexity of the combinational test generation still remains.

The complexity of designs is still growing and the demand for fault models and tests for faults much closer to physical defects than the stuck-fault model grows as well. To generate tests for all those faults exhaustively will be prohibitive. Hence, the system under test has to be partitioned into modules in an intelligent way to perform test generation as a divide-and-conquer procedure. Test generation will then be a two step procedure, first to generate a module test to cover the faults assumed (local test) and then to calculate the test access of this module from primary inputs/outputs of the entire circuit (global access).

Scan designs partition the circuit by separating state register from combinational logic but the test access times are unacceptable. Another way to get test access to modules without undesired test time overhead is via multiplexers [WH96]. This method has no restrictions on the partitioning scheme but, in practice, modules are taken by construction, i.e. functionally well defined units with proven local test sets. The chip is switched to several test modes in which the modules are tested at function and at speed.

In both schemes there is no need for exploitation of behavioral information. The decision where to place test access points is given by some structural indications solely. In any case, a lot of effort is spent to reduce area overhead for the test equipment on the chip.

On the other hand it can be shown that calculations of stimulus and response on behavioral descriptions is much faster than on structural descriptions. Consequently, before implementing additional test access hardware it should be figured out which tests or test sequences can be switched to and from an embedded module by its environment. FunTestIC, like shown, is able to derive such frame sets and by this to support the test access derivation. Prerequisite is a circuit partitioned by functional criteria. Thinking of the design flow, e.g. [ArGr93], this will always be the case if HDL behavioral descriptions are employed which are decomposed in the behavioral domain first. After having skipped to the structural domain all further synthesis steps must not blur the functional module structure. The loss of bounds in many cases results from optimizations during synthesis. Thus, optimizations have to be done with care taking into account the bounds of functional modules. If this functional decomposition is to be changed by a latter synthesis step this has to be back annotated.

Concluding, automatic or nearly automatic test generation for complex circuits and systems postulates corresponding behavioral and structural descriptions. Synthesis tool development tends to this goal. As shown, exploiting the behavioral view makes lot of the problems of structural test generation much easier to solve.

References

[Ak78] Akers, S.B.: Functional Testing with Binary Decision Diagrams; *Proc. 8th Symp. Fault Tol. Comp.*, pp. 75-82, 1978.

[ArGr93] Armstrong, J.R., Gray, F.G.,: *Structured Logic Design with VHDL*; Prentice-Hall, ISBN 0-13-885206-1, 1993.

[BaAr86] Barclay, D.S., Armstrong, J.R.: A Heuristic Chip-Level Test Generation Algorithm; *23rd Design Automation Conference*, pp. 257-262, 1986.

[BaAr86] Barclay, D.S., Armstrong, J.R.: A Heuristic Chip-Level Test Generation Algorithm; *23rd Design Automation Conference*, pp. 257-262, 1986.

[BBH94] Bashford, S., Bieker, U., Harking, B., Leupers, R., Marwedel, P., Neumann, A., Voggenauer, A., Voggenauer, D.: *The MIMOLA language 4.1*; University Dortmund, Informatik 12, 1994.

[BiKa96] Bieker, U., Kaibel, M., Marwedel, P. , Geisselhardt, W.: Hierarchical Test of Embedded Processors by Self-Test Programs;

[BM95] Bieker, U., Marwedel, P.: Retargetable self-test program generation using constraint logic programming; *32nd Design Automation Conference*, 1995.

[BM89] Bhattacharya, D., Murray, B.T., Hayes, J.P.: *High-Level Test Generation for VLSI*; IEEE Computer, pp. 16-24, 1989.

[Bra95] Brakel, G. van: *Test Pattern Generation for Delay Faults*; Doctoral Thesis, Universiteit Twente, Dept. Electr. Eng., 1995.

[Cha77] Chan, A.Y.: *Easy-to-use Signature Analysis Accurately Troubleshoots Complex Logic Circuits*; Hewlett-Packard Journal, Vol. 28, pp. 9-14, 1977.

[ChAr94] Cho, C.H., Armstrong, J.R.: B-algorithm: A Behavioral Test Generation Algorithm; *Proc. ITC*, pp.. 968-979, 1994.

[Che88] Cheng, W.: The Back Algorithm for Sequential Test Generation; *Proceedings International Conference on Computer Design*, pp. 66-69, 1988.

[CP90] Cheng, W., Patel, J.H.: PROOFS: A Super Fault Simulator for Sequential Circuits; *Proc. Europ. Design Autom. Conf.*, pp.475-479, 1990.

[EK95] Emshoff, B., Kaibel, M.: Einfuehrung Binaerer Entscheidungsgraphen auf hoher Beschreibungsebene in VHDL; *Proc. in German 7. EIS Workshop*, pp. 129-132, 1995.

[GK91a] Gouders, N., Kaibel, R.: Advanced Techniques for Sequential Test Generation; *Proc. ETC*, pp. 293-300, 1991.

[GK91b] Gouders, N., Kaibel, R.: PARIS: A parallel pattern fault simulator for synchronous sequential circuits; *Proc. ICCD*, pp. 542-545, 1991.

[Gl94] Gläser, U.: Mehrebenen - *Testgenerierung für Synchrone Schaltwerke*; Doctoral Thesis in German, Univ. Duisburg, Electrical Eng. Dept., 1994.

[GM93] Geisselhardt, W., Mojtahedi, M.: New methods for parallel pattern fast fault simulation for synchronous sequential circuits. *Proc. ICCAD*, 1993.

[GMo89] Geisselhardt, W., Mohrs, W., Moeller, U.: *FUNTEST - Functional Test generation for VLSI-Circuits and Systems*; Microelectronic Reliability, Vol. 29, No. 3, pp. 357-364, 1989.

[Go91] Gouders, N.: *Methoden zur deterministischen Testgenerierung für synchrone Schaltwerke*; Doctoral Thesis in German, Univ. Duisburg, Electrical Eng. Dept., 1991.

[Goe81] Goel, P.: *An implicit Enumeration Algorithm to Generate Tests for Combinational Logic Circuits*; IEEE Trans. Comp., Vol. C-30, pp. 215-222, March 1981.

[GSC91] Giambiasi, N., Santucci, J.F., Courbis, A.L., Pla, V.: Test Pattern Generation for Behavioral Descriptions in VHDL; *Proc. EuroVHDL'91*, Stockholm, pp. 228-235, 1991.

[GuSt96] Gulbins, M., Straube, B.: Applying Behavioral Level Test Generation to High-Level Design Validation, *Proc. of ED&TC*, pp. 613, 1996

[HLM83] Hassan, S.Z., Lu, D.J., McCluskey, E.J.: Parallel Signature Analyzers - Detection Capability and Extensions; *26th IEEE COMPCON*, pp. 440-445, 1983.

[He68] Hennie, F.C.: *Finite-State Models for Logical Machines*; John Wiley & Sons, 1968

[HVT91] Huemmer, H.-D., Veit, H., Toepfer, H.: Functional Tests for Hardware Derived from VHDL Description; *Proc. CHDL*, pp. 433-445, 1991

[IEEE92] Design Automation Standards Subcommittee of the IEEE: Draft standard VHDL language reference manual; IEEE Standards Department, 1992.

[Jo79] Johnson, W.A.: Behavioral-Level Test Development; *Proceedings 16th Design Automation Conference*, pp. 171-179, 1979.

[KE96] Kaibel, M., Emshoff, B., Geisselhardt, W.: Investigations on High-Level Control for Gate-Level ATPG, *Proc. European Test Workshop*, Montpellier, pp. 192-201, 1996.

[LM82] Levendel, Y.H., Menon, P.R.: *Test Generation Algorithms for Computer Hardware Description Languages*; IEEE Trans. on Computers, Vol. C-31, No. 7, pp. 577-588, 1982.

[LePa91] Lee, J., Patel, J.H.: An Architectural Level Test Generator for a Hierarchical Design Environment; *Proc. 21th Symp. on Fault-Tolerant Comp.*, pp. 44-51, 1991.

[Li84] Lin, T.: The S-Algorithm: A Promising Solution for Systematic Functional Test Generation; Proc. *Int. Conference on Computer-Aided Design*, pp. 134-136, 1984.

[MG95] Marwedel, P., Goossens, G. editors: *Code Generation for Embedded Processors*; Kluwer Academic Publishers, 1995.

[Mo94] Mojtahedi, M.: *Methoden zur Beschleunigung der automatischen Testmustergenerierung und Fehlersimulation für synchrone sequentielle Schaltungen*; Doctoral Thesis in German, Univ. Duisburg, Electrical Eng. Dept., 1994.

[My79] Myers, G.J.: *The Art of Software Testing*; John Wiley & Sons, 1979.

[NJA89] O'Neill, M.D., Jani, D.D., Cho, C.H., Armstrong, J.R.: BTG: A Behavioral Test Generator; *Proc. CHDL*, pp. 347-361, 1989.

[NJA90] O'Neill, M.D., Jani, D.D., Cho, C.H., Armstrong, J.R.: BTG: A Behavioral Test Generator; *Proc. CHDL*, pp. 347-361, 1990.

[Roth66] Roth, J.P.: *Diagnosis of Automata Failures. A Calculus and a Model*; IBM Jounal of Research and Development, V.9, No. 2, 1966.

[Sa89] Sarfert, T.M., Markgraf, R., Trischler, E., Schulz, M.H.: Hierarchical Test Pattern Generation Based on High-Level Primitives; *Proc. IEEE Int. Test Conf.*, pp. 1016-1026, Sept. 1989.

[Sprc92] SPARC International Inc.: The SPARC Architecture Manual; Version 8, Prentice Hall, 1992

[Ve92] Veit, H.H.: *A Contribution to the Generation of Tests for Digital Circuits Described by Their Behavior*; Doctoral Thesis , Univ. Duisburg, Electrical Eng. Dept., 1992

[ViAb92] Vishakantaiah, P., Abraham, J., Abadir, M.: Automatic Test Knowledge Extraction from VHDL (ATKET); *29th ACM/IEEE Design Automation Conference*, pp. 273-278, 1992.

[WH96] Huemmer, H.-D., Wiemers, T., Geisselhardt, W., Splettstoesser, W.: TAMUX: Test Access via Multiplexers with minimum area overhead and at speed testability; *3rd International Test Synthesis Workshop*, Sta. Barbara, CA, May 6 - 8, 1996

[WiPa82] Williams, T.W., Parker, K.S.: *Design for Testability: a Survey*; IEEE Trans. Comp., Vol. C-13, No. 1, pp. 2-15, 1982.

[YoSi96] Yount, Ch.R., Siewiorek, D. P.: *A Methodology for the Rapid Injection of Transient HardwareErrors*; IEEE Trans. on Comp., No.8, pp.881-891, 1996.

11 BEHAVIORAL FAULT SIMULATION

Jean-François Santucci
Paul Bisgambiglia
Dominique Federici

11.1 Introduction

Due to the ever-increasing complexity of VLSI circuits, the use of VHDL [Vhd87] behavioral descriptions in the fields of test generation and fault simulation becomes advised. In order to understand this evolution and the context in which it is efficient, the increasing complexity of VLSI circuits must be considered. To better cope with the complexity of existing and future systems, higher abstraction levels must be taken into account. Furthermore, the only knowledge about the device being tested may come from data sheets or signal measurements. In this case the only way to generate test patterns is behavioral testing. Today, such a task is done manually and one of the main interests in Behavioral Test Pattern Generation (BTPG) is to automate it. Another point to consider is that the test generation process is an integral part of the design process and its implementation must begin during the behavioral design phase.

Motivated by these facts, various BTPG methods have been proposed in the recent past [Bar87,One90,Nor89,Hum91,Cou92,Stra93]. Most of these BTPG methods are based on a deterministic generation process using a formal fault model : each sequence is constructed in order to detect a given fault belonging to a Fault List determined from a Behavioral Fault Model.

Despite the use of heuristic criteria [Che91,San92,San93], to speed up the search for test sequences, these methods remain very time-consuming because of the

261

J.C. López et al. (eds.), Advanced Techniques for Embedded Systems Design and Test, 261-284.
© 1998 *Kluwer Academic Publishers.*

deterministic nature of the BTPG processes and the lack of Behavioral Fault Simulation (BFS) methods.

In order to facilitate the definition of a powerful BFS method, it is essential to explicitly represent the concepts involved in behavioral descriptions. We have therefore defined an internal model which highlights on the one hand the sequential and concurrent aspects and on the other hand the separation and interaction between the control flow and the data flow.

The list of considered behavioral fault hypothesis is derived using two steps : (i) definition of an exhaustive fault model which collects all possible fault hypothesis in terms of the incorrect «functioning» of modeling items ; (ii) selection of a sub-set of conventional faults [Gho91].

Our approach for defining a BFS method leans on the resolution of the three following sub-problems in order to deal with high-level behavioral descriptions :

- 1- determination of the high level basic elements of the internal model which will have a key role in BFS.

- 2- definition of how lists of behavioral faults are propagated through the previous basic elements. The simulation of a basic element requires deducing the fault list at the basic element outputs from the input faults lists.

- 3- development of an overall algorithm allowing to propagate fault lists through the complex data and control structures involved in the internal model.

The BFS method has been implemented in C++ language and first results have been obtained.

In the next section of the chapter we deal with the main concepts involved in behavioral test pattern generation. Section 3 presents the motivation in behavioral testing. The interest of using of a behavioral fault simulator coupling with a deterministic and a pseudo-random behavioral test pattern generator is detailed in section 4. Our approach for performing BFS is given in section 5. Section 6 deals with implementation and results. Perspectives are overviewed in a concluding part.

11.2. What is Behavioral Testing

11.2.1 Concepts of Behavioral Testing

Testing is a major activity in CAD (Computer Aided Design) systems for digital circuits. For that reason, CAD systems include ATPG (Automatic Test Pattern Generation) tools. ATPG techniques can be divided into two families depending on the model type of the circuit under test : structural or behavioral.

The structural view or « white box » represents a potential structure of the studied system as an interconnection of basic elements. This structure is qualified as potential because there is not necessarily a complete mapping between the model and the actual circuit. The interconnected items model physical components which can be in turn modeled according either a structural or behavioral view. It should be pointed out that the item belonging to the lowest level of a structural description must be described according to a behavioral point of view.

The behavioral view or « black box » represents the studied system by defining the behavior of its outputs according to values applied on its inputs without knowledge of its internal organization. A behavioral model can be defined by means of two kinds of representation : alphanumeric representations or graph-based representations. Alphanumeric models are textual representations which describe the system behavior by means of declarative or procedural programming languages. Graph representations (binary decision diagrams [Ake78a] used in [Ake78b,Aba85,Cha86], transformation graphs [Lai83], Petri Nets [Pet81] used in [Olc91,Scho85]) are characterized by an interconnection of basic elements for which a behavior is predefined as context-free. It should be noted that this structure is not a potential structure of the modeled system as is the case in the structural view.

Whatever the view by which the circuit is modeled, an abstraction level is taken into account.

We have to point out that abstraction levels are not implied by a given view : the boolean level does not have to imply a structural view nor does system level have to imply a behavioral view.

The structural test refers to a test pattern generation methodology based on a structural model of the circuit under test.

The behavioral test refers to a test pattern generation methodology based on a behavioral model of the circuit under test.

11.2.2 Fault Modeling

The aim of the generation process is to define patterns to apply on primary inputs in order to test eventual physical defects. These defects can be detected only if they induce an irregular behavior called a fault. The fault effect (or error) is measured by a difference between the state of the fault-free model (reference model) and the state of the faulty model (model in which a fault hypotheses is injected). Let us point out that the definition of test patterns usually leans on the definition of fault hypothesis depending on both the abstraction level and the view from which the circuit is modeled. A fault hypothesis is :

- either an hypothesis of an incorrect behavior of an item belonging to the model, but considered context-free

- or an hypothesis of modification of the initial global description by adding, suppressing or combining basic items, without modifying the predefined behavior of these items.

It should be pointed out that the modification of the global description of a model usually lead to take into account too large a number of possibilities. Thus this kind of fault hypothesis is not usually considered. In a similar way, overly complex faults concerning the modification of a basic item behavior will also be excluded.

11.2.3 Behavioral Test Pattern Generation

Behavioral Test Pattern Generation concerns the development of softwares dedicated to the determination of set of patterns allowing behavioral faults to be detected. Behavioral Fault Models are defined using the approach described in sub-section 2.2. Recently a set of work[Bar87,One90,Nor89,Hum91,Cou92,Stra93] has been developed in this area.

11.3 Motivation in Behavioral Testing

11.3.1 Main advantages of Behavioral Testing

Recent advances in VLSI technology have had a great impact on testing. For the last ten years, research work has been oriented towards Behavioral Test Pattern Generation. in contrast with conventional structural approaches [Kir87,Schu87,Fuj87]. In order to understand this evolution and the context in which it is efficient, the increasing complexity of the VLSI circuits must be considered. It is now well known that high abstraction levels must be taken into account when dealing with Design and Test of Digital Systems. One very important point is that the test generation process is an integral part of the design process and its implementation must begin during the behavioral design phase. The potential advantages include early estimates of the coverage rate of physical failures prior to the synthesis as well as a faster fault simulation.

11.3.2 Confidence in Behavioral Testing

The test quality depends on both the abstraction level and the view taken into account. Obviously, for a given abstraction level, generating from a structural view is better than generating from a behavioral model in terms of physical failures coverage.

However a set of work has proven that some measure of confidence could be highlighted for Behavioral Testing.

A first solution to address the validity of behavioral testing consists in performing a fault simulation of behavioral test sequences on a structural model described at a lower level of abstraction (i.e. the gate level single stuck-at fault model). This approach has been carried out in [Gho91] and encouraging results for Behavioral Testing have been pointed out.

A second solution is to establish a correlation between behavioral fault hypothesis and physical failures.

11.4 Behavioral Test Pattern Generation using Fault Simulation

11.4.1 Overview of a General BTPG System

Figure 11.1 shows a system overview allowing to perform an efficient BTPG. The inputs of such a system is a VHDL behavioral description of the circuit for which the BTPG has to be performed and the list of target faults. The outputs of the system comprises of the generated test patterns, three lists containing the detected faults, the redundant faults and the aborted faults.

A detected fault is a fault for which a test pattern exists. A redundant fault is a fault which can not be detected by any test pattern. An aborted fault is a fault for which no solution has been found using the deterministic BTPG within the backtrack limit ; obviously, this is a limit of the BTPG software. The BTPG process will require an important time consuming without this backtrack limit.

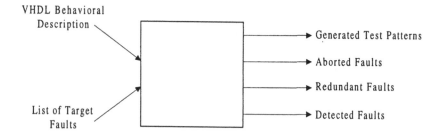

Figure 11.1. BTPG System

In order to speed up the behavioral test generation process, an efficient BTPG has to involve two distinct phases:

- Phase 1 : a Pseudo-Random T.P.G (PRTPG) ; the goal is to randomly generate test sequences using a process which is not time-consuming. The approach is successful if the random generation process allows the target fault list to be drastically reduced.

- Phase 2 : a deterministic T.P.G.

The BFS approach presented in section 6 has to be used for grading the generated test patterns in both phases of the ATPG process.

11.4.2 Behavioral Pseudo-random Test Generation using BFS.

The pseudo-random test generation process is an inexpensive way to generate test sequences. By definition, in a pseudo-random method, tests are generated through a quite random process guided by two types of information : the number of patterns which have to be generated and a criterion that patterns have to fulfill.

In order to estimate the length of the test sequences we have been concerned by complexity-based metrics [Cou95] stemming from the Software Engineering field. Complexity-based metrics have been first proposed to evaluate the complexity of software in order not to design software which cannot be easily tested. An interesting application of the complexity metrics to the testing activity was developed in [Mca76]. In this work, McCabe defined the cyclomatic number of the control part of a software. He proved that this number is similar to a path coverage and represents the minimum number of test data to be generated to test the control part.

In order to evaluate the «quality» of the random generated test patterns the inexpensive way is to use behavioral fault simulation.

The reasons for choosing a PRTPG as test generation strategy in phase 1 can be summarized as follows :

- PRTPG is straightforward.

- The number of new faults detected per test is initially large. Thus after a few trials, the fault coverage achieved is sufficiently high to cause a significant decrease in the fault simulation cost per pattern.

- The deficiencies of the PRTPG are that the number of new faults detected per pattern decreases and that the fault simulation effort for the evaluation of those random patterns, which do not detect any additional fault, is wasted. So as soon as an acceptable fault coverage is obtained or the percentage of new detected faults is very low, the random generation of test patterns ends. Only faults which are not detected by the random test sequences are subjected to a deterministic generation process.

The approach involving BFS during a random generation process is summarized in figure 11.2.

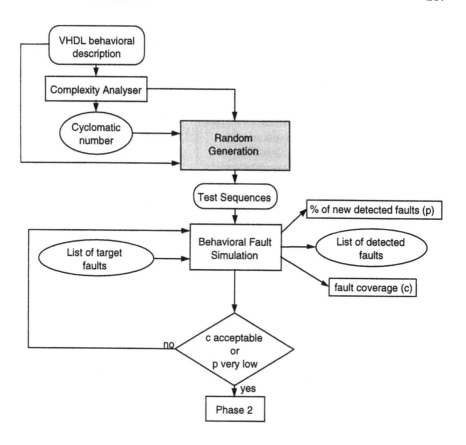

Figure 11.2. Phase one of the BTPG system

11.4.3 Deterministic BTPG using BFS.

The list of remaining faults not yet detected during phase 1 are concerned with phase 2. A deterministic behavioral test pattern generator derived from the path sensitization family introduced in section 2.3 is used during phase 2 in order to generate test patterns.

However instead of deterministically generate test patterns for each fault, once again we use the BFS in order to improve the generation process.

Our approach for phase 2 is illustrated in Figure 11.3.

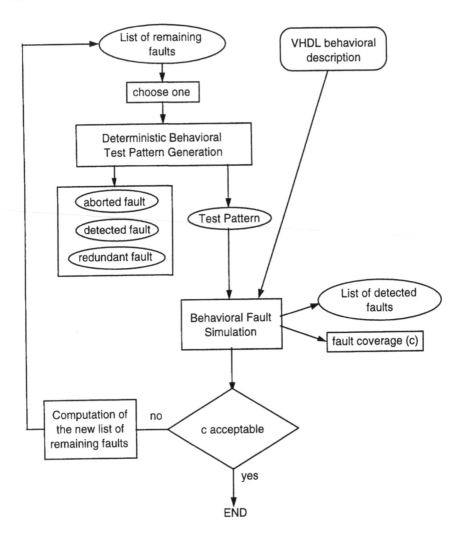

Figure 11.3. Phase two of the BTPG system

11.5 An approach for Behavioral Fault Simulation

11.5.1 Petri-Net Modeling of VHDL Behavioral Descriptions

In order to describe the circuit behavior the most commonly used language is VHDL [Vhd87]. A behavioral view allows a circuit to be described independently of its internal structure by defining how its outputs react when values are applied to its inputs. An internal model is associated with each behavioral description. This model must satisfy two requirements : it should be easy to handle, and able to highlight the concepts linked to behavioral descriptions for a high abstraction level. In order to satisfy the first requirement the definition of the internal model is based on graph concepts. In order to address the second requirement the internal model has been decomposed into two parts, the separation and the interaction of which are explicit : an control part representing the sequencing of operations and pointing out the different execution paths of the description and a data part pointing out the data handled and the operations which transform or test them.

The control part is modeled by a structural graph stemming from the formalism of Petri Nets [Pet81] involving two types of nodes : places and transitions. Structured sub-graphs are predefined for modeling the scanning phase, process activation, control structures (sequential, repetitive or selective) involved in VHDL descriptions.

The Data part is modeled using a graph with two types of nodes : data nodes and operation nodes. There are two classes of data nodes representing input/output signals or internal variables. Each data node has a value attribute allowing its current value to be memorized. Furthermore, in case of signal data nodes an additional attribute is used in order to memorize the previous value (required during the simulation cycles execution). For a signal called S, this attribute is noted S_{old}.

Lastly the interaction of the Control part and the Data part is achieved by associating an operation node with a place node. Two links are involved between the place and the operation node : an activation link of the operation node and an end report link.

Figure 11.4 represents the different entities involved in our modeling scheme.

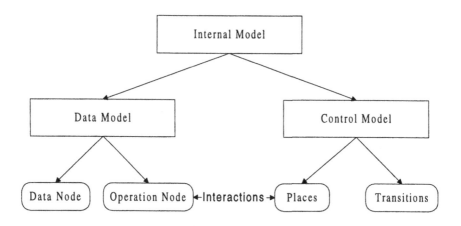

Figure 11.4. Basic entities of the internal modeling

This modeling scheme is illustrated by Figures 11.5 and 11.6. Figure 11.5 gives the VHDL behavioral description of a 8 bit register.

```
Entity Register IS
Port (DI : IN vlbit_1d(1 TO 8) ;
            STRB, DS1, NDS2 : IN vlbit ;
            DO : OUT vlbit_1d(1 TO 8)) ;
END Register
Architecture behavior of Register IS
SIGNAL reg : vlbi_1d(1 TO 8);
SIGNAL enbld : vlbit ;
BEGIN
  strobe: PROCESS (STRB)
  begin
    If (STRB =1) then reg <= DI;
    End If;
  END PROCESS ;
  enable : PROCESS (DS1,NDS2)
  begin
    enlbd <= DS1 AND NOT (NDS2);
  END PROCESS;
  output : PROCESS (reg,enbld)
  begin
    If (enbld=1) then DO<=reg
    Else DO <= 11111111;
    end If
  END PROCESS;
END Behavior
```

Figure 11.5. VHDL behavioral description

The corresponding internal model is described in Figure 11.3. The data part is not represented in that figure.

Figure 11.7 gives more details concerning the first process of the 8-bit register (Process STROBE) ; we have represented in this Figure both the control and data parts. We have to point out that our modeling scheme involves pure Petri Nets. The dynamical aspect of the interaction of the two models is supported by tokens. When a token arrives in a place, the associated operation is performed. In case of a decision node, the result associated with the end report link is used in order to select the next transition to fire.

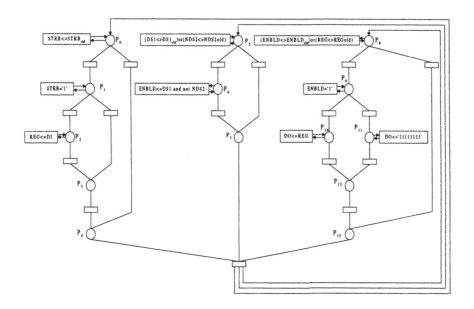

Figure 11.6. Internal Model of the 8-bit Register behavioral description

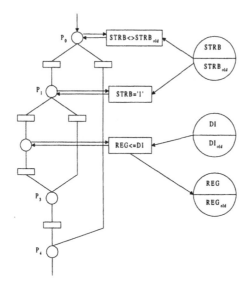

Figure 11.7. Process Strobe's internal modeling

11.5.2 Behavioral Fault Modeling Scheme

A fault modeling scheme allows to derive a set of fault hypothesis on the previously described model. Fault hypothesis are defined according to the elements involved in the internal model of the circuit under test. In order to have a behavioral fault model for which some measures of confidence are provided, we select fault hypothesis according to the fault model proposed in [Gho91].

The selected fault hypothesis are classified into the following three groups. The examples given in order to illustrate fault hypothesis belonging to the different groups are issued from the process Strobe (see section 5.1) :

Stuck-at fault on an element of the data model:

. **F1** : Stuck-at of a data node. It corresponds to the stuck-at of a signal (or a variable).

Example : REG_0 means that the value of the data node REG is stuck at 0.

Stuck-at fault of an element of the control model:

. **F2** : Faults on control elements. $F2_t$ (resp. $F2_f$) : the true branch (resp. the false branch) of the control element is always selected whatever the resulting value set up on the end reporting value.

Example : Let «a» be the selective place corresponding to the operation node STRB=«1». The fault $F2_{at}$ (resp. $F2_{af}$) is expressed by : whatever the results of the test «STRB=1?» is, the branch STRB=1 (resp. STRB=0) is selected.

Stuck-at fault of an element of the interaction of the control and data models:

. **F3** : Faults on the interaction data-control elements. F3 is the statement jump.

Example : $F3_1$ is expressed by the jump of the assignment number 1 (REG<=DI).

We have to point out that the focus of the paper is not the definition of a behavioral fault model. Furthermore the selected fault model may be modified but the BFS method we detail in the next section will still remain valid.

11.5.3 Overall Algorithm for BFS

The BFS technique described above is based on the gate level deductive fault simulation [Arm72]. We will describe in this sub-section the different phases of our method.

First, we give the two restrictions concerning the behavioral descriptions we want to test :

• Unique fault hypothesis.

• A given signal can not be assigned in two different processes.

The principle of our method is composed in two phases (see figure 11.8) : the Fault Free Simulation (FFS), and the Fault List Propagation (FLP).

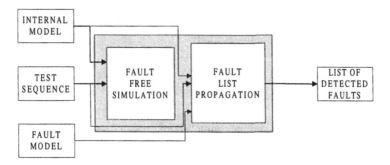

Figure 11.8. Overview of our methodology.

11.5.3.1 The Fault Free Simulation

The input data are : a test pattern, an internal model representing the behavioral description of our circuit. The goal of this step is to perform a fault free simulation of the circuit for the given test sequence in order to obtain two results :

- the fault free values of the primary output(s) of the system.

- the number of performed simulation cycles (nb_cycle)

11.5.3.2 The Fault List Propagation

In this sub-section, we define how fault lists are propagated through the internal model during the simulation.

Figure 11.9 represents the flow chart corresponding to the FLP step.

A first loop (boxes 1,2,3,4,10) allows the propagation of fault lists for the previously determined number of cycles.

A second loop (boxes 3,4,5,6,7,8,9) allows to achieve the fault list propagation through each process involved in the VHDL description.

The propagation through one process is performed using the procedure SIMULATION (box 8).

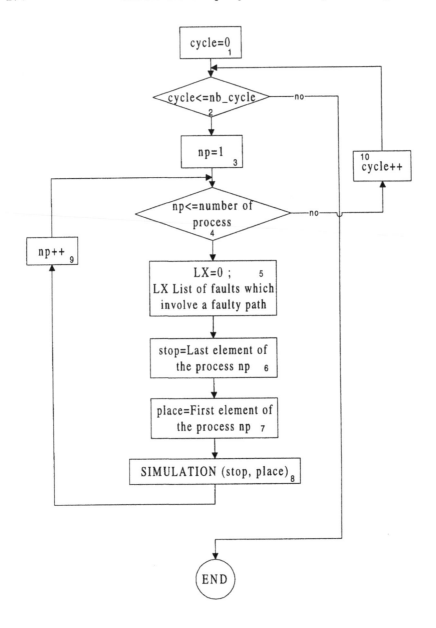

Figure 11.9. Flow chart of the FLP step

The flow chart of the procedure SIMULATION is given in Figure 11.10.

This procedure searches through the control model of the considered process until the last place is encountered (box 1). According to the type of the current place (boxes 2 and 5), two different rules are applied:

"If the place corresponds to an assignment operation, the following rule is used (box 6)"

We consider the assignment : DATA1 <= E(DATAi) with 0<i<n and E a function.

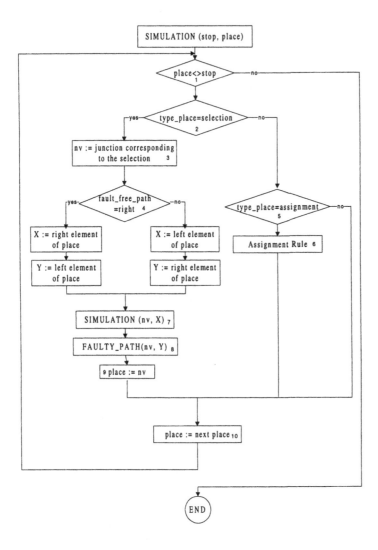

Figure 11.10. Flow chart of the procedure SIMULATION

We give the rule allowing to compute the fault list L_{DATA1} for an input vector T1 through an assignment element. This input vector allows to assign a value to the variable DATA1 using the algebraic or Boolean operation E(DATAi) where DATAi are represented by data nodes in the data part of the internal model.

$L_{DATA1} = \cup L_{DATAi} \cup Ljump-L$

- L is the empty set if the value of the attribute old of the data node DATA1 is equal from the value computed using E(DATAi) given the input vector T1. In the other case, L is the fault list LDATA1 computed the last cycle.

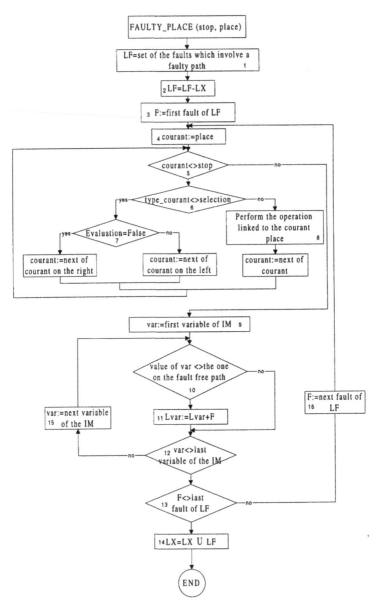

Figure 11.11. Flow chart of the procedure FAULTY_PATH

- Ljump is the empty set if the value of the attribute old of the data node DATA1 is equal to the value computed using E(DATAi) given the input vector T1. In the other case, Ljump is equal to the singleton $\{F3_{assignment}\}$.

- $\cup L_{DATAi}$ is the union of the fault lists associated to the data nodes DATAi these lists are composed of the faults on the input signals of E(DATAi) which are locally observable on the output DATA1.

> *"If the place corresponds to a selection operation, the rule is expressed by the procedure FAULTY_PATH (box 8)"*

The flow chart of this procedure is given in Figure 11.11.

The principle allowing to perform the fault list propagation is the following :

For each fault F belonging to the Fault List LF (box 1), we compute the values of the signals involved in the corresponding faulty path (boxes 6,7, 8).

The obtained values are compared with the ones found during the FFS (box 10)

If one of these values are different, the fault F is added to the list Lvar (list of faults locally detected on the signal var) (see box 11).

The detected faults are the ones associated with the primary output signals.

11.5.4 Pedagogical Example

This sub-section is dedicated to the description of the application of our technique on an example. We will describe a step by step fault simulation. The considered circuit is a 8-bit register, its VHDL Behavioral description and its internal model are given in the section 5.1.

The list of faults being simulated (derived from the fault model presented in 5.2) is the following :

$L=\{$ $STRB_0$, $STRB_1$, DI_0 , DI_1 , $DS1_0$, $DS1_1$, $NDS2_0$, $NDS2_1$, DO_0 , DO_1 , REG_0 , REG_1 , $ENBLD_0$, $ENBLD_1$, $F2_{P0V}$, $F2_{P0F}$, $F2_{P1V}$, $F2_{P1F}$, $F2_{P5V}$, $F2_{P5F}$, $F2_{P8V}$, $F2_{P8F}$, $F2_{P9V}$, $F2_{P9F}$, $F3_{P2}$, $F3_{P6}$, $F3_{P6}$, $F3_{P11}$ $\}$.

The test sequence we choose is : (*DI=00000000* , *STRB=1* , *DS1=1* , *NDS2=0*)

11.5.4.1 The Fault Free Simulation

Table 11.1 resume the fault free simulation phase.

The two results which will drive the fault propagation phase are :

- The fault free value of the primary output : *DO=00000000*.

- The number of cycles executed : *nb_cycle=2*.

	Initial State	Test Sequence (Bold line)	State after the first cycle	Final State
REG	00000000	00000000	00000000	00000000
DI	00000000	**00000000**	00000000	00000000
STRB	0	**1**	1	1
ENBLD	0	0	1	1
DS1	0	**1**	1	1
NDS2	0	**0**	0	0
DO	00000000	11111111	11111111	*00000000*

Table 11.1. Fault Free Simulation

11.5.4.2 The Fault List Propagation

We describe in this sub-section the propagation of the fault hypothesis during the step by step simulation.

Nbcycle=0

This is the initialization phase (all the processes are activated).

Process STROBE

$L_{P1V} = \{$ $STRB_1$, $F2_{P1V}$ $\}$

These faults are lost because there is no difference of the signals values between the fault free path and the faulty path.

Process ENABLE

$L_{ENBLD} = \{$ $DS1_1$, $ENBLD_1$ $\}$

Process OUTPUT

$L_{P9V} = \{$ $ENBLD_1$, $F2_{P9V}\}$

We find these faults on DO.

$L_{DO} = \{DO_0$, $F3_{P11}$ $\} \cup \{$ $ENBLD_1$, $F2_{P9V}\}$

We have in the fault list L_{DO} all the faults which involve a change on the value of DO with the one we found during FFS.

Nbcycle=1

This is the first cycle of the execution phase. Strobe and Enable are activated , Output is suspended .

Process STROBE

$L_{POF} = \{$ $STRB_0$, $STRB_1$, $F2_{POF}$ $\}$ and $L_{P1F} = \{$ $F2_{P1F}$ $\}$ are not locally detected.

$L_{REG} = \{$ REG_1 , DI_1 $\}$

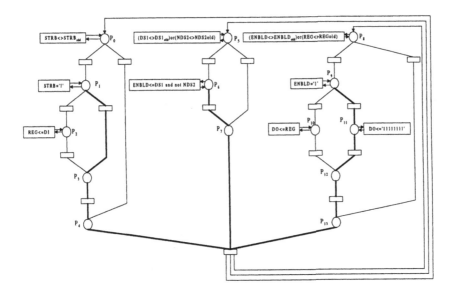

Figure 11.12. Initialization cycle

Process ENABLE

$L_{P5F}=\{$ DS1_0 , DS1_1 , F2$_{P5F}$ }(these faults are locally detected on ENBLD and added to L_{ENBLD}).

$L_{ENBLD}=\{$ ENBLD$_0$, NDS2$_1$, F3$_{P6}$ } $\cup$ { DS1_0 , DS1_1 , F2$_{P5F}$ }- L_{ENBLD}.

Process OUTPUT

$L_{P8V}=\{$ DS1_1 , F2$_{P8V}$ }(these faults involve a faulty path).

Since F2$_{P8V}$ has no influence on ENBLD it is not locally detected (no change on the signal DO)

DS1_1 leads to ENBLD=1 ; as a result DO<=REG is performed under this fault hypothesis ; hence DS1_1 is locally detected on DO.

$L_{DO}=\{$ DO$_0$, F3$_{P11}$, ENBLD$_1$, F2$_{P9V}$, DS1_1 }

Nbcycle=2

Process OUTPUT

$L_{P8F}=\{$ ENBLD$_1$, ENBLD$_0$, F3$_{P6}$, NDS2$_1$, DS1_1 , DS1_0 , F2$_{P5F}$, F2$_{P8F}$ }

$L_{P9F}=\{$ F2$_{P9F}$ }, this fault is locally observable on DO.

$L_{DO}=\{$ DO$_1$, F3$_{P10}$, REG$_1$, DI$_1$ }

$L_{DO\ FINAL}= L_{DO} \cup L_{P9F} \cup (L_{P8F}-L_{DO\ cycle\ 1(car\ DO\ change)})$

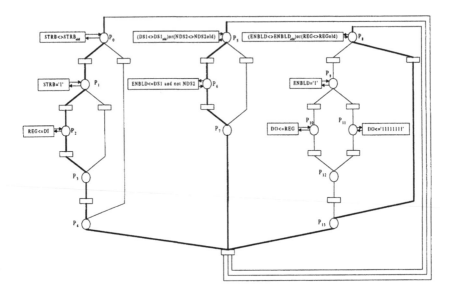

Figure 11.13. First cycle of the execution phase

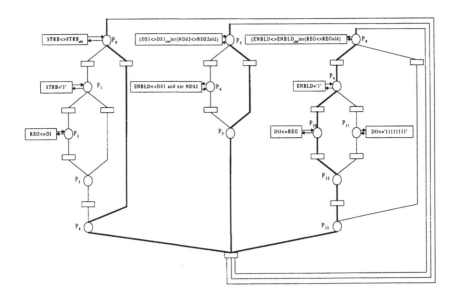

Figure 11.14. Second cycle of the execution phase

11.5.5 Implementation and First Results

The BFS method presented in section 4 was coded in C++ language on a Sparc 5 Workstation according to object-oriented paradigms [Stro89].

The BFS software system is decomposed into three parts :

- generation of an ASCII file from a VHDL description ; this file allows to explicitly point out the control and data parts involved in the VHDL description (figure 11.15).

- generation of C++ instances allowing to describe both the internal and fault modeling schemes ; these instances are obtained from two C++ Classes files : control/data models Classes and fault hypotheses Classes (figure 11.15).

- the general algorithm allowing to perform the BFS is implemented in C++ language ; the fault lists propagation rules are executed using the messages sending notion (figure 11.16).

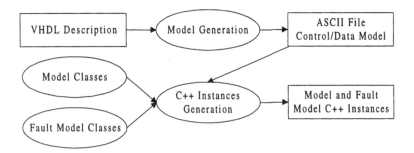

Figure 11.15. Generation of the C++ instances involved in our software

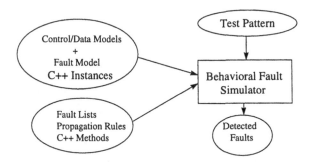

Figure 11.16. overview of our BFS software

Experiments were performed on experimental behavioral descriptions writen in VHDL. These experiments were conducted in order to validate the Behavioral Fault Simulation algorithm propose in section 5.3.

We give the final results of the BFS software applied to the VHDL description presented in section 5.1. The test pattern used for the experiment is the one presented in the pedagogical example (sub-section 5.4).

Test Pattern : (DI=00000000 , STRB=1 , DS1=1 , NDS2=0)

Type of Fault	Considered signal or place	Stuck at :
F_2	P_8	F
F_1	ENBLD	0
F_2	P_5	F
F_1	DS1	0
F_3	P_6	
F_1	NDS2	1
F_2	P_9	F
F_3	P_{10}	
F_1	DO	11111111
F_1	REG	11111111
F_1	DI	11111111

Table 11.2. Results of our BFS Software

11.6 Perspectives

In this chapter, we have presented a new approach for Behavioral Fault Simulation. By using an internal modeling allowing the main features of behavioral descriptions to be highlighted, we have been able to define an efficient BFS method. By transposing the fault model proposed in [Gho91] on the elements of the internal model, we make sure that the conclusion reported in [Gho91] about the confidence of the fault model are still characterized our corresponding fault model.

The approach we propose consists into three steps : (i) definition of basic elements allowing to guide the fault lists propagation, (ii) elaboration of simple rules allowing fault lists to be propagated through these previous basic elements, (iii) definition of an overall algorithm allowing the fault list propagation through the internal model using the previous rules.

This approach has been illustrated on an example by detailing each main point of the algorithm.

We have also presented a new approach for Behavioral Test Pattern Generation for digital circuits using Behavioral Fault Simulation.

We have defined the specifications of the BFS software using an Object Oriented Design Tool and implemented the BFS software in C++ language. We have presented the first results obtained using a set of experimental behavioral VHDL descriptions.

Our future investigations are summarized in the following :

- we envision to use a deterministic Behavioral Test Pattern Generator [Nor89,Hum91,Cou92,Stra93,Che91,San92] in order to obtain test patterns for faults remaining not detected after the pseudo-random test pattern generation.

- the experimental scheme has to be complemented with real VHDL behavioral descriptions.

Furthermore, the validation of the behavioral fault model is not addressed in this chapter. However, it is a crucial open problem which must be solved in order to see people from industry becoming confident in Behavioral Testing, even if it has been shown in [You96] that there was a correspondence between fault at the RT level and fault at the gate level.

References

[Aba85] M.S. Abadir, H.K. Reghbati, "Functinal test generation for LSI circuits described by binary decision diagrams", *International Test Conference*, pp. 483-492 (1985).

[Ake78a] S.B. Akers, *"Binary decision diagram"*, IEEE Trans. on Computers, Vol. C-27 (6), pp. 509-516 (1978).

[Ake78b] S.B. Akers, "Functional testing with binary decision diagrams", *8th Fault Tolerant Computing Symposium*, pp. 82-92 (1978).

[Arm72] D.B. Armstrong, *"A Deductive Method for Simulating Fault in Logic Circuits"*, IEEE Trans. on Computers, Vol C-21, N°5, pp. 464-471, May 1972.

[Bar86] D.S. Barclay, J.R. Armstrong, "A chip-level Test Generation Algorithm", *23th IEEE/ACM Design Automation Conf. 1986*, pp.257-261.

[Cha86] H.P. Chang, W.A. Rogers, J.A. Abraham, "Structured functional level test generation using binary decision diagrams", *IEEE International Test Conference*, pp. 97-104 (1986).

[Che91] C. H. Chen, C. Wu, D. G. Saab, "BETA : Behavioral Testability Analysis", *IEEE ICCAD'91*, Santa Clara, pp. 202-205, 1991.

[Cou92] A.L. Courbis, J.F. Santucci, N. Giambiasi, "Automatic Test Pattern Generation for Digital Circuits", *1st IEEE Asian Symposium*, Hiroshima, pp. 112-118, 1992.

[Cou95] A.L Courbis, J.F. Santucci, "Pseudo-Random Behavioral ATPG", *Great Lakes VLSI Symposium*, pp. 192-195, 1995.

[Fuj87] H. Fujiwara, T. Shimono, "On the acceleration of test generation algorithms", *International Test Conference*, pp. 1016-1026 (1987).

[Gho91] S. Ghosh, T.J. Chakraborty, *"On behavior fault modeling for digital designs"*, Journal of Electronic Testing : Theory and Applications, Vol. 2, pp. 135-151, 1991.

[Hum91] H.D. Hummer, H. Veit, H. Toepfer, "Functional Tests for hardware derived from VHDL description", *CHDL 91*, pp. 433-445, 1991.

[Kir87] T. Kirkland, M.R. Mercer, "A topological search algorithm for ATPG", *24th Design Automation Conference*, pp. 502-507 (1987).

[Lai83] K.W. Lai, D.P. Siewiorek, "Functional testing of digital systems", *20th Design Automation Conference*, pp. 207-213 (1983).

[Mca76] T.J. Mcabe, *"A Complexity Measure"*, IEEE Trans. Software Engineering, N°2, pp. 308-320, 1976.

[Nor89] F.E. Norrod, "An automatic test generation algorithm for hardware description language", *26th Design Automation Conference*, pp.76-85, July 1989.

[Olc91]S. Olcoz, J.M. Colom, "Petri Net based synthesis of VHDL programs", *Euro-VHDL'91*, Stockholm (September 1991).

[One90] M.D. Oneill, D.D. Jani, C.H. Cho, J.R. Armstrong, "BTG : a Behavioral Test Generator", *9th Computer Hardware Description Languages and their Application*, IFIP, pp.347-361, 1990.

[Pet81]J.L. Peterson, *"Petri Net Theory and the Modeling of Systems"*, Prenctice-Hall, Inc., Englewood Cliffs, New York (1981).

[San92] J. F. Santucci, G. Dray, N. Giambiasi, M. Boumédine, "Methodology to reduce computational cost of behavioral test pattern generation using testability measures", *IEEE/ACM 29th DAC*, pp. 267-272, 1992.

[San93]J. F. Santucci, A. L. Courbis, N. Giambiasi, "Speed up Behavioral Test Pattern Generation using an Heuristic Criterion", *IEEE/ACM 30th DAC*, pp. 92-96, 1993.

[Scho85]P.D. Schotts, T.W. Pratt, "Hierarchical modeling of software systems with timed Petri Nets", *International workshop on Timed Petri Nets*, Turin, Italy, pp. 32-39 (June 1985).

[Schu87]M.H. Schulz, E. Trischler, T.M. Sarfert, "Socrates : A highly efficient automatic test pattern generation system", *International Test Conference*, pp. 1016-1026 (1987).

[Stra93] B. Straube, M. Gulbins, E. Fordran, "An approach to Hierarchical test Generation at Algorithmic Level", *IEEE Workshop on hierarchical test Generation* , Blackburg, Virginia, USA, 8-11 august 93.

[Stro89]B. Stroustrup, *"Le langage C++"*, Interedition, 1989.

[Vhd87] *VHDL Language reference Manual*, IEEE Standard 1076, 1987

[You96] C.R. Yount, D.P. Siewiorek, *"A Methodology for the Rapid Injection of Transient Hardware Errors"*, IEEE Trans.on Computers, pp. 881-891, August 1996.

INDEX